Signal Timing Under Saturated Conditions

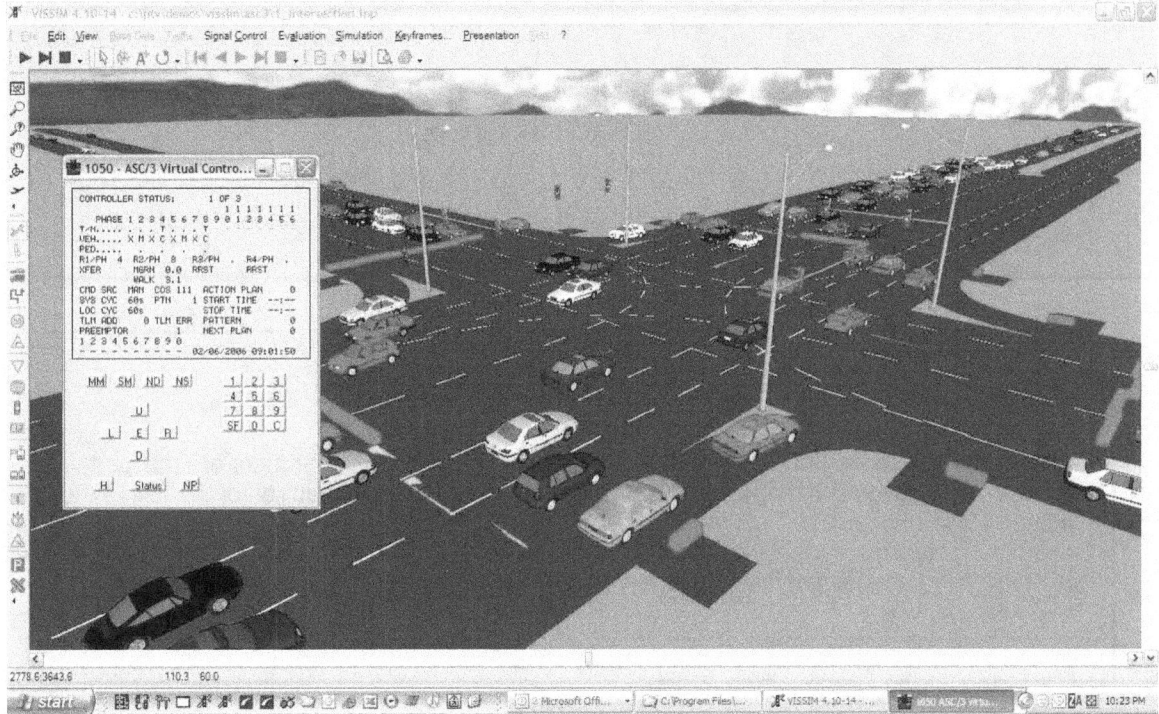

FINAL REPORT

by
Booz|Allen|Hamilton
Iteris
University of Arizona
Econolite

November 2008

U.S. Department
of Transportation
Federal
Highway
Administration

NOTICE

This document is disseminated under the sponsorship of the U.S. Department of Transportation in the interest of information exchange. The U.S. Government assumes no liability for the use of the information contained in this document. This report does not constitute a standard, specification, or regulation. The U.S. Government does not endorse products or manufacturers. Trademarks or manufacturers' names appear in this report only because they are considered essential to the objective of the document.

QUALITY ASSURANCE STATEMENT

The Federal Highway Administration (FHWA) provides high-quality information to serve Government, industry, and the public in a manner that promotes public understanding. Standards and policies are used to ensure and maximize the quality, objectivity, utility, and integrity of its information. FHWA periodically reviews quality issues and adjusts its programs and processes to ensure continuous quality improvement.

1. Report No. FHWA-HOP-09-008	2. Government Accession No.	3. Recipient's Catalog No.
4. Title and Subtitle Signal Timing Under Saturated Conditions		5. Report Date: November 2008
		6. Performing Organization Code
7. Author(s) Principal Investigator: Richard W. Denney, Jr., P.E. Co-Authors: Larry Head, Ph.D., Kevin Spencer		8. Performing Organization Report No. Project
9. Performing Organization Name and Address Booz Allen Sub-consultants: Iteris, Inc.; University of Arizona; Econolite		10. Work Unit No. (TRAIS)
		11. Contract or Grant No. Contract No. DTFH61-06-D-00006
12. Sponsoring Agency Name and Address U.S. Department of Transportation Federal Highway Administration Office of Operations 1200 New Jersey Ave, SE Washington, DC 20590		13. Type of Report and Period Covered Final Report October 2006 to September 2008
		14. Sponsoring Agency Code HOP

15. Supplementary Notes

Eddie Curtis (Eddie.Curtis@fhwa.dot.gov) was the Technical Representative for the Federal Highway Administration. Agency experts and consultants provided interviews in support of identifying the state of the practice, including:

- Gerard de Camp, independent consultant
- Steven Click, Tennessee Tech University (formerly with North Carolina DOT)
- Woody Hood, Maryland State Highway Administration
- Eric Nelson, Harris County (Texas) Department of Transportation
- Gary Pietrowicz, Road Commission of Oakland County (Michigan)
- Ziad Sabra, Sabra-Wang and Associates
- Bill Shao, Los Angeles Department of Transportation

16. Abstract

This report provides guidance to practitioners on effective strategies to mitigate the effects of congestion at signalized intersections. The scope is limited to single intersections and does not address network-level strategies. The strategies are defined in terms of their underlying objective. Under congested conditions, traditional objectives and performance measures shift from progression and minimizing delay to maximizing throughput and managing queues. Experts were interviewed to identify the strategies and tactics they used to address congested intersections, a discussion of their methods is presented. Select strategies were studied further, particularly the belief that longer cycles are more efficient, and the effects of buses on signal timing in grid networks. The research revealed that long cycle lengths may not be more efficient at intersections where long queues starve turn lanes. In grid networks, the cycle length that is just long enough to reliably serve busses from a near-side stop was found through simulation to prevent the development of residual queues.

17. Key Words Signalized Intersections, Traffic Signal Timing, Congestion, Cycle Lengths, Throughput, Queue Management		18. Distribution Statement No Restrictions. This document is available to the public	
19. Security Classification. Unclassified	20. Security Classification. (of this page) Unclassified	21. No. of Pages 76	22. Price N/A

Signal Timing Under Saturated Conditions   November 2008

Table of Contents

Acknowledgments 3
Guidance 4
   About This Document 4
   Objectives 5
      Traditional Objectives 5
         Minimizing Cycle Failures 5
         Maximizing Performance Measures 5
      Maximizing Throughput 6
      Queue Management 8
   Strategies 8
      Guidance from Experts 8
         Common Evaluation Themes 9
         Throughput Strategies 9
         Queue Management Strategies 11
         Demons 11
         Controller Features 12
      Cycle Length and Green Time 12
         Buses 13
I. Literature Review 14
   Overview 14
   Annotated Bibliography 16
      Foundational Research 16
      Optimization Model Research 17
      Network Metering 18
      Genetic Algorithms 18
      MCH1542 B Installation Guide for MOVA 19
   References 20
   Overview of NCHRP 3-66 21
      References for NCHRP 3-66 Discussion 25
II. The State of the Practice 26
   Categorizing Congestion Based on Operational Objectives 26
   Example Applications 31
      Square Lake Road at Telegraph Road, Oakland County, Michigan 31
      Sunrise Boulevard, Rancho Cordova, California 32
      North Carolina State Highway 54 and I-40, Durham, NC 34
      Bandera Road and Guilbeau Road, San Antonio, Texas 35
   State-of-the-Practice Interviews 37
      Definition of Saturation 38
      Motivating Conditions 38
      Strategies Taken, and Their Objectives 39
      Demons 42
      Tactics and Controller Features 42
      Desired Controller Features 45
III. Evaluation of Strategies 47
   Long Green Times and Cycles 47
      Introduction 47
      Study Site 49
      Field Data 50
      Analysis of Field Data 55

    Field Data Conclusions and Further Discussion 58
    Simulation 60
    Discussion and Conclusion 67
Cycle Length and Bus Capacity *68*
    Introduction 68
    Modeled Operation 70
        Signal Operation 70
        Demand 70
    Simulation 70
    Simulation Results 72
        Cycle Length 72
    Conclusion 74

Acknowledgments

Much of the work reported herein was inspired by discussion held in the Transportation Research Board Committee on Traffic Signal Systems, which has held several conference sessions on the subjected of signal timing in the congested regime. Particularly, the presentations at the 2006 summer meeting, held in Woods Hole, Massachusetts, provided an excellent overview and framework for further work. The committee also provided useful feedback during a series of reports by the researchers on the progress of this effort.

The simulation work for the San Antonio scenario for looking at the effect of buses was performed by Kevin Spencer at the University of Arizona.

The foundation of the work is based on interviews conducted with a number of recognized signal timing experts, and the researchers in particular acknowledge and express gratitude for their willingness to participate in the lively discussions that led to the description and understanding of the state of the practice. Those experts include:
- Gerard de Camp, independent consultant
- Steven Click, Tennessee Tech University (formerly with North Carolina DOT)
- Woody Hood, Maryland State Highway Administration
- Eric Nelson, Harris County (Texas) Department of Transportation
- Gary Pietrowicz, Road Commission of Oakland County (Michigan)
- Ziad Sabra, Sabra-Wang and Associates
- Bill Shao, Los Angeles Department of Transportation

Guidance

About This Document

This document is organized to provide guidance as a summary of the material that follows. The subsequent chapters provide detailed documentation of the work performed that lead to this guidance. It should be noted that some guidance is supported by results from interviews with a panel of experts, some is based on research conducted as part of this project, and some is a synthesis of practice recommended by the study team. The text will attempt to make these sources clear so that the reader can reasonably assess the reliability of the guidance.

The project was organized in three parts. The first part (Chapter I) was a literature review on the general subject of signal timing in congested conditions and on the characterization of saturation. The literature review continued with a discussion of the NCHRP 3-66 project, which was anticipated to relate to this project to a greater extent than actually proved to be the case. The literature review was expanded slightly to include a review of work done on the subject of cycle length, long green times, and saturation flow. Because that work is directly related to the work of the first section of Chapter III, it is included there.

The second part of the project (Chapter II) identifies methods used by acknowledged experts to address congested conditions using signal timing. To collect data and information through interviews, the research team developed a framework for categorizing methods by their underlying objective; this is a dominant theme of the Guidance section. The interviews identified a range of methods that might be used in congested conditions. Some are in the form of principles to understand, and some are specific approaches to signal timing. All of the proposed methods are limited to solutions that can be implemented at the intersection level. Network-level solutions are being explored in the related NCHRP 3-90 project, which is underway at the time of this writing, and which will incorporate the findings of this project.

The third part of this project (Chapter III) involved simulation and field studies to evaluate some of the methods that emerged from the review of the state of the practice. Not all methods could be researched within the context of this project, but focus was given to those issues that are often misunderstood or that have an extraordinary impact on operations in the opinion of the research team. These studies are documented in Chapter III.

This Guidance section will start with a discussion of objective functions as the basis for decided what tools to implement in the congested case. Understanding how the methods relate to the objectives will help avoid self-defeating interactions. The authors recommend that all practitioners consider their objectives when faced with a congested scenario, before deciding what to do.

Following the discussion of objectives, this Guidance chapter will present the strategies that emerged from the interviews and from the research that was conducted.

Objectives

Practitioners are trained in methods, and those methods are designed to achieve a specific goal (not always perfectly). For example, one might time traffic signals to maximize progression, minimize delay, minimize cycle failures, minimize fuel consumption, maximize throughput, manage the location and size of queues, and so on. Not all signal timing methods are characterized in terms of their objective, with the result that practitioners sometime employ methods inappropriate to the situation at hand. It is the recommendation of the researchers that practitioners carefully consider their objectives. To assist with this recommendation, the researchers have characterized objectives in the following sections.

Traditional Objectives

Minimizing Cycle Failures
This objective seeks to minimize the frequency that a traffic signal does not serve all the waiting cars during the green period.

The Yale Bureau of Highway Traffic, which was the original traffic engineering study program at the post-graduate level in the U.S., promoted a technique for determining cycle length and green times based on mean arrivals. Once mean arrivals were known, an assumption that arrivals were random could be used as the basis for estimating the number of arrivals that would only be exceeded a small percentage of the time (say, 5%). The Poisson Distribution was used to make this estimation, and this method has come to be known as the Poisson Method. Because it seeks to provide green times long enough to accommodate nearly all arriving flows, it demonstrates the objective of minimizing cycle failures.

Cycle failures occur when the green time is not long enough to serve all the cars that were waiting at the start of green. In and of themselves, cycle failures are not evidence of congestion. But when the remaining queue at the end of green, which we call the *residual queue*, grows over a series of cycles, most experts will agree that the conditions have become congested.

Minimizing cycle failures is not the same thing as minimizing residual queuing. As volumes increase, the 95th percentile arrival will become too large to accommodate in a cycle, and the method will no longer converge on a reasonable solution. Also, the method looks only at expected arrivals, not at actual arrivals, and thus uses a cycle that is longer than needed much of the time. Actuation at a non-coordinated signal might overcome that problem to some extent, if the detectors are carefully placed and the actuated control logic so optimized (see NCHRP 3-66 for more guidance on this issue).

Thus, the Poisson Method, and any other method with the objective of minimizing cycle failures, is only appropriate in light traffic conditions where the motorist expectation is to be served on the next available green.

Maximizing Performance Measures
In this objective, the methods used seek to increase some positive performance measure or decrease a negative performance measure to the extent possible.

To address the limitations of the Poisson Method, researchers started to develop tools that would achieve more attainable objectives in heavier traffic conditions. These tools characterize performance measures using analytical or empirical simulation models, and adjust timing systematically to optimize those performance measures.

The performance measures are typically one of two possibilities:
- Progression, where coordination timings are adjusted to maximize the opportunity to drive through successive signals on a green light.
- Delay and stop in some combination, as a general performance measure.

Both of these have problems in the presence of growing residual queues. Progression assumes that vehicles can move unimpeded down the street as long as the signal timings are appropriate, and usually ignore the possibility of a residual queue providing that impedance.

Delay is also difficult to characterize in the congested regime. Queues form and grow when demand exceeds capacity, and the delay caused by the queue depends on its length. Thus, longer queues mean more delay. If queues are growing, the delay becomes a function of how long one collects the data, rather than a measure of performance that can be optimized.

Thus, delay minimization and progression, as tools for practical signal timing, are limited to moderate traffic conditions free of growing residual queues.

In the research reported in Chapter III, the researchers define "demand exceeds capacity" more carefully as that point at which increases in the offered load at the intersection can no longer be served. That serves as the definition of saturation, and it is visible in the field by growing residual queues. Thus, the presence of repeated residual queues is the symptom that should encourage the practitioner to move to a different objective.

Maximizing Throughput

In this work, the researchers develop the objective of maximizing throughput, or the number of cars actually served by the intersection, with respect to the cars presented to the intersection (the load).

A cycle failure causes a residual queue, which adds to the demand to be served on the next cycle, in addition to new arrivals. Even if the new arrivals are few enough to be served by the next green period, the residual queue from the previous cycle might cause a cycle failure by consuming green time needed by the freshly arriving flow. Thus, normal variation in traffic that causes occasional cycle failures makes it more difficult for the signal to recover without a change in timing strategy. If the variation in demand provides sufficient reductions in demand such that the residual queue can just clear before growing again, then we might call this condition *intermittent congestion*. Any increase in average demand will prevent this recovery. Periods of several failed cycles followed by a cycle or two that clear, followed by more failed cycles provide a useful indication that the capacity of the current operation has been reached, and the operation is crossing into the congested regime.

The residual queue prevents progression, so progression is no longer a meaningful objective as demand approaches capacity at key intersections in the system (except perhaps for carrying traffic away from a bottleneck intersection). Also, delay becomes difficult to define. Intersection delay can be characterized by the area between the arrival flow curve and the departure flow curve, as shown in the Figure G-1.

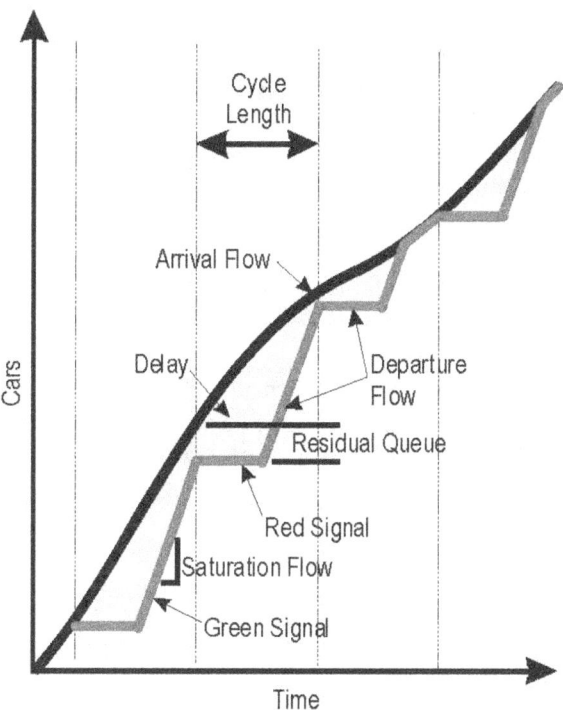

Figure G-1. Delay resulting from arrival and departure flows at an intersection.

(It should be noted that the figure suggests that the departure flow maintains saturation flow either until the queue dissipates or the light turns red, and this research will show that this is an assumption that does not hold for long green times when the approach traffic contains turning traffic that cannot reach turn lanes. The shape in the figure are idealized for clarity.)

When demand exceeds capacity and the residual queues no longer clear, the delay grows as the queue grows and the arrival and departure flow curves in the intersection diverge.

Another way to define this point is to compare the demand of the intersection to the throughput by counting cars going into the intersection versus cars coming out of the intersection. As the queue grows, the cars coming out reach a maximum that is constrained by the available green time, while the arrives cars are not constrained. Figure 3-10, later in the this document, shows a typical relationship between the demand and throughput, where demand is characterized as *offered load* and throughput as *served load*. In this work, we have defined capacity as the point where throughput no longer increases with increased demand. *Demand* and *throughput* are traffic engineering terms that are similar to the optimization terms of offered and served loads, and we use these terms interchangeably in this report.

When an intersection has residual queuing that is just beginning to grow over time, the maximum throughput of the intersection has been reached.

The practitioner should then first attempt to improve the maximum throughput to the extent possible by implementing one of a range of signal timing strategies (or by avoiding one of several signal timing mistakes). These strategies and mistakes are the subject of this research and the next section will describe them.

Maximizing throughput has the aim of minimizing queue formation to prevent congestion. At some point, however, no revision to the signal timing will increase maximum throughput, and the queues will continue to grow until demand diminishes. The practitioner's objectives will need to shift again, this time to one of queue management.

Queue Management

The queue management objective seeks to use signal timing to allow queues to form where they do the least damage.

When growing residual queues can no longer be relieved by maximizing throughput, then the practitioner's only choice is to arrange the operation of signals within a network to prevent the queues from multiplying the problem by creating capacity constraints of their own. This usually means constraining capacity upstream from a bottleneck at locations where queue storage will not cause network gridlock, or where safety problems might occur as a result of queue formation.

If throughput maximization is a curative approach, then queue management can be considered a palliative approach with the objective of treating symptoms rather than seeking a cure.

Strategies

In this project, the researchers adopted two approaches to identifying effective methods for optimizing signal timing in congested conditions to satisfy the objectives of optimizing throughput or managing queues. The first of these was to form a panel of acknowledged expert practitioners. The members of the panel were interviewed to identify examples from their experience that illustrated techniques that they used. These techniques were then evaluated to determine their underlying objective (which was mixed in some cases). Some were used as the basis for further consideration. In all cases, the experts knew what they were attempting, even if they did not articulate it, and identifying their objective helps to explain their approach. We recommend that practitioners consider their objective when addressing congestion at an intersection. All the experts first attempted to cure the congestion by maximizing throughput, and when that was not adequate they employed queue management strategies.

Next the researchers attempted to evaluate several approaches identified during the interviews that warranted further research. Some of these approaches can be adequately explained and recommended without further research, but some issues and observations were inconsistent with the observed beliefs of some practitioners, and therefore warranted further research. Those issues were evaluated using simulation and, in one case, field studies.

Guidance from Experts

Project researchers conducted a series of interviews with leading practitioners from around the country (See Chapter II, The State of the Practice for details). The interviews were not structured, and the intent was to let the experts lead the discussion as a means of identifying underling principles, objectives, and priorities. Each interviewee also provided a case study, and many of these are documented in Chapter II. This section will summarize those interviews in the form of guidance for practitioners.

Common Evaluation Themes

Experts were not generally interested in defining the point of saturation in a precise or scientifically elegant way, but were rather interested in conditions that would justify a change in strategy or more detailed adjustment to operation on the ground. When pressed, the experts generally agreed with a definition of saturation as when maximum throughput is exceeded, or when residual queues grow. These were reported as being triggers for action, and some tightened their action level to growing queues that caused more general network problems. When discussing methods for evaluating congested locations, some common themes emerged. These included:

- Inadequacy of existing evaluation tools
- Inadequacy of existing optimization tools
- The necessity for performing evaluations on the ground by direct observation.
- The necessity to devote sufficient observation to fully understand the problem and the effect of candidate solutions.

Evaluation tools that were generally considered to be inadequate included capacity methods (as a means of identifying maximum throughput) or any computer-based models short of some (and only some) microscopic simulation. For the experts, the determination of congested was based on a growing queue problem, and that was determined solely by direct observation.

The experts also generally did not use signal timing optimization tools to help solve a congestion problem. Most reported that those methods seemed insensitive to even the existence of growing residual queues, let alone the input of that condition into the optimization objective. This impression is supported by a cursory review of available tools, which based signal timing on delay or some other performance measure, or on balancing the saturation level of the various movements at the intersection. No existing popular tools could be identified by the experts that optimized specifically to maximize throughput or to manage queues. The recommendation to practitioners is to avoid depending on signal timing optimization tools to provide solutions for situations where growing residual queues indicate that the objective should shift to maximizing throughput.

In all the case studies reported here, the experts invested considerable time in observing, understanding, and addressing the problem. Resources are constrained in most agencies, and this devotion of attention reflects the priority given by experts in addressing points of congestion. In many cases, the expert spent days of observation before formulating a plan, and then days of implementation and operation to refine and evaluate the solution. The recommendation for practitioners emerging from these examples is to invest the time the most where the congestion is the worst, because that is where limited resources may do the greatest good. Another recommendation is to commit the most experienced experts available in the organization to observe and address the worst problems.

Throughput Strategies

A number of general techniques presented by the experts could be categorized as strategies aimed at maximizing throughput. These include:

- Work back from downstream bottleneck
- Run close intersections on single controller
- Improve lane utilization
- Lag heavy left turns
- Find the right cycle
- Serve heavy movements more than once in a cycle
- Consider effect of buses
- Minimize effect of pedestrian movements
- Seek all possible available green time, even if esoteric signal controller features are required
- Consider congested and uncongested movements separately
- Prevent actuated short greens

These can be broadly divided into three categories:
- Making best use of the physical space provided in the intersection
- Making best use of green time in the cycle
- Minimizing the negative effects of other influences.

Several of these were studied in greater detail in later project phases, including running close intersections on a single controller, considering the effect of buses, and, most important of all in the opinion of the researchers, finding the right cycle. These will be summarized further down in this Guidance section and reported in more detail in later chapters.

All of these methods can be readily implemented in existing signal control equipment, though a more straightforward phase reservice capability was mentioned as desirable. Phase reservice is a feature that allows one phase to be served more than once in a cycle. Chapter II shows an example of how this might be done with a very simple eight-phase controller. Phase reservice serves both a throughput objective and a queue management objective. By serving congested movements twice in a cycle, uncongested movements will consume less of the effective cycle without violating required minimum green periods or other constraints. The example in Chapter II shows how minor movements, by being served only once for every two times the congested phases are served, improve the percentage of the cycle devoted to the congested movements. This is the single most effective strategy for relieving one congested movement in cases where the uncongested movements are constrained to longer green times than needed by cars or to exclusive phase protection.

A closely related principle was making use of any feature of the controller that could serve to increase the percentage of the cycle devoted to the congested movements. Several of the experts agreed that they were prepared to use even extremely esoteric combinations of controller features to that end, though one expert countered that attitude by avoiding techniques that could not be understood or maintained by maintenance technicians.

Most of the experts warned that actuation could defeat the objective of maximum throughput. They did not mention that sluggish extension timings could cause greens to be extended for traffic of lower flow rather than serving higher flows on other movements, though the experts generally supported sharp actuated timing to minimize extension, and other experts worked in the context of coordination when addressing congested signals. They did, however, make extensive use of actuation features to maximize the amount of the cycle that would serve maximum flow over the stop line. They also warned of congested movements being cut short because of a single sluggish vehicle such as a truck, with the observation that even one untimely gap out could significantly increase a residual queue such that it would take many cycles to recover.

Queue Management Strategies

When throughput cannot be maximized sufficiently to control the queues at a congested intersection, the experts suggested a range of techniques that are aimed at managing the resulting queues to minimize the propagation of congestion within the network. They include:

- Reduce minor splits to encourage diversion (usually doesn't work)
- Run close intersections on single controller
- Balance queue for conflicting approaches
- Prevent queue from multiplying congestion
- Meter traffic into bottlenecks
- Prevent backing queues into bottlenecks

Some of these also appeared as throughput strategies, such as running close intersections on a single controller. This approach should be given priority when carrying queues through two intersections is too important for minimizing congestion to be relegated to a low-priority controller function such as coordination. Coordination can be overridden by most controller functions, including preemption, pedestrian and minimum greens, and (in some cases) maximum greens. Also, coordination implies cycle-based patterns that will usually change during the day, resulting in transitions during which coordination may break down temporarily. When two intersections are run from a single controller, however, the relationship between them can be built into the phasing design and therefore maintained even during high-priority controller functions such as preemption. Diamond interchanges provide an example that will be explored further down.

Demons

The experts also identified conditions or situations that work against optimizing for throughput or queue management. These include:

- Longer and longer vehicle clearances (especially red clearances)
- Longer pedestrian clearances
- Too much dependence on detection

The clearance issue is especially timely. The 2003 edition of the Manual on Uniform Traffic Control Devices has increased the time required for pedestrian crossings to allow pedestrians to cross the entire traveled way, instead of just to the center of the far lane. The FHWA has proposed for the next edition of the MUTCD to reduce the assumed walking speed used for calculating the pedestrian clearance time from 4 feet/second to 3.5 feet/second. If adopted, this will change pedestrian crossing times yet again.

More subtle, though, are increases in red clearance. Many agencies have been ratcheting up red clearances for a variety of reasons. These are justified on the basis of having only a small effect on signal timing, but with the expectation of a larger effect on safety. The small effect on signal timing is based on an often false assumption, however.

The assumption is that the clearance is provided by reducing the green time for the phase being cleared.

This is not the case, however, at intersections where one movement is congested. The other movements will have been reduced to an arbitrary minimum green, either constrained by the agency's policy on minimum greens or by pedestrian clearance times. Thus, the green times of the uncongested movements cannot be further shortened—they are as short as they are allowed to be. An increase in red times for these movements do not take away from the green time for these movements, but rather accumulate around the signal cycle to the congested phases, where green time that can be shortened is available. Thus, even a modest increase in clearance of only, say, one second in each phase might result in shortening a congested green by four seconds, which further might reduce throughput by, say, 8%, assuming a 50-second green in a 100-second cycle. Thus change in clearances, which was assumed to be insignificant, can result in a significant reduction in green time just where the green time is most needed. The true consequences of these clearance interval decisions should be considered before those decisions are made and implemented.

Controller Features

The experts interviewed for the project described a range of signal controller features that they employed to address some aspect of maximizing throughput or managing queues. These included:

- Phase Reservice, with the general recommendation to serve congested movements more often in the cycle, or alternatively to serve movements that are too lightly traveled to deserve their minimum greens on alternate cycles.
- Minimize Ped Effects
    - Let Peds override forceoffs, or
    - Use additional Max in lieu of forceoffs by programming max time for the normal coordinated green time and accommodating ped times within the coordination timings. With max not inhibited, the green will be routinely short because the phase will max out, except when a ped call is being served, in which case the ped intervals will override the max, all within the context of the coordinated plan.
- Use special logic, such as terminating the side street if remaining cars are right turns that can be served during complementary left turn
- Simultaneous Gaps—don't let phase extend again if detection is received during green rest
- Dual Entry so that late-arriving cars on the opposing minor approach won't cause a delayed start of green
- Volume-Density variable gap to prevent actuated short greens, such as those that might be caused by a truck accelerating from stop too slowly.

Cycle Length and Green Time

The experts noted that flow seemed to decrease as green times extended beyond about 30 seconds, even when fed by an arbitrarily long queue. Project researchers explored this issue by conducting a field test at a site in Northern Virginia, as reported in Chapter III. That work has led to several conclusions:

- Saturation flow does not decline in lanes unaffected by cars leaving the lane, but

- Lanes adjacent to turn lanes do see a significant reduction in saturation flow as turners depart the through lane for the turn lanes. At the study site, a 12% turning volume resulted in a significant reduction in throughput for the through movement. The reduction was sufficient to reduce the overall throughput of very long cycles, as indicated by simulation studies based on the field site.

The guidance for practitioners is that on approaches with turn lanes, the green time and cycle should be kept short enough to serve only vehicles queued up to the upstream entrance to those turn lanes. That way, the green time serves only through cars, with the assumption that turning cars will be served during other phases (as with left turns) or in lanes that can move with other phases (as in both left and right turns). In many locations, this green time may result in unacceptably short minor movements or in violations of required minimum intervals, and in those cases, the shortest cycle that does not cause those problems may provide the greatest throughput.

Practitioners are encouraged to abandon the commonly held belief that long cycles are naturally more efficient because lost time is reduced as a percentage of the cycle. This effect was not confirmed, and other effects were found to be much more dominant.

Buses

One effect that emerged from the experts was that cycle length was related to bus capacity. The researchers tested this concept using simulation of three intersections along a downtown one-way street. The simulation was formulated to eliminate the effect of cycle length on progression, and the upstream intersection metered flow into the network by providing less green time for the main street than downstream intersections. Thus, the simulation was sensitive only to the effects of buses.

The simulation shows that changes in bus volumes did not significantly affect the throughput of the street. But the size of the residual queue was significantly affected by changes in the cycle length. The conclusion is that the interaction of buses and the cycle length can have a significant effect on congested flow, and that it is that interaction with cycle length, not bus volumes per se, that have the effect.

The cycle length that performed the best was the shortest cycle that could reliably serve at least one bus arriving during that cycle. The cycle that performed the worst was the longest cycle that could not reliably serve more than one bus. Thus, for buses that stop on a congested approach and that require a typical 20-second loading and unloading time, the best cycle is about 40 seconds and the worst cycle is about 80 seconds. The cycles below 40 and the cycle above 80 were noticeably worse in their effect on the number of cars in the network (and therefore the total delay) than the cycles from 40 to 70 seconds. The guidance is therefore to use the shortest cycle on congested approach that carry significant bus demand, and that have a stop on the approach, that will reliably serve at least one bus per cycle. In the simulation, that was shown to be about twice the normal bus loading and unloading time. The cycles from that value up to about 50% longer provide the best operation in this scenario.

I. Literature Review

Overview

Considering the impact of oversaturated signals on motorist delay and network performance, surprisingly little research has been conducted that leads to specific and implementable methods to recommend to practitioners. Much of the literature that does exist is devoted to quantifying the queue that results when the green time is insufficient to serve the arrival demand. Most of this research has not been cited in this section. Work cited here provides some attempt at responding to the saturation problem rather than merely quantifying traffic flow in a signal-imposed queue.

In the early 1960's, Gazis studied oversaturated conditions and wrote the earliest papers found in the literature. He assumed that the *time duration* of the oversaturated period, the demand curve, and the service curve were empirically known and objective was to minimize the area between these two curves, which is a measure of the total delay. When the arrival demand cannot be served by the service curve, the two curves diverge and the amount of unserved queue grows. Many attempts have been made to characterize delay while considering the unserved queue. For example, a deterministic departure process leads to the following equation, as reported by Rouphail (1995):

$$d = \frac{(c-g)}{2c\left(1-\frac{q}{S}\right)}\left[(c-g)+\frac{2}{q}Q_o+\frac{1}{S}\left(1+\frac{I}{1-\frac{q}{S}}\right)\right]$$

where

d = delay

c = cycle length

g = effective green time

q = arrival flow

S = saturation flow

Q_o = expected queue at end of green (residual queue)

I = Poisson index of dispersion (a measure of arrival stochasticity)

The problem with this and other delay models is that they require knowledge of the residual queue size, Q_o. With a variable residual queue in the right-hand side of the equation, and delay on the left, delay cannot be calculated as a steady-state measure, and the equation has no closed-form solution without an assumption of the time period, and without disregarding the delay of the vehicles that remain in the residual queue at the end of that time period.

After Gazis's publication, a substantial effort to identify a closed-form analytical estimate of the residual queue ensued. These efforts failed in the general case, leading to the approach of developing computationally convenient delay models that could be evaluated at demand levels exceeding estimates of capacity. Modeling predictable values of delay when demand exceeds capacity was a compromise with reality, and the commitment to using delay as an optimization objective override the inability calculate delay in those conditions.

In Gazis's work, the resulting optimal control policy was constructed by dividing the oversaturated period into two stages, in the first stage, the maximum green is allocated to the phase with highest flow rate and the minimum green time is allocated to the other phase. In the second stage the maximum and minimum green times are switched, to oscillate between residual queues on the two approaches with the objective of balancing them. Thus, the objective of the approach was to balance queue formation. The optimal cycle length and switching point from stage to stage were calculated based on the traffic volumes and duration.

This basic approach was modified to include queue capacity constraints by Michalopoulos, Stephanopolos, and May (1978). More recently Chang and Lin (2000) developed a discrete state space version of Gazis's model. The advantage of the discrete model is that the switch point is always at the end of a cycle.

Daganzo (1995) introduced a Cell Transmission Model (CTM) to capture the dynamics of traffic flow. In the CTM the entire traffic network is divided into small cells with each cell length defined as the distance a vehicle can travel at free flow in one time step. Lo and others (2004) applied the CTM to oversaturated traffic networks with the objective to minimize the total delay of the whole network. Using the CTM approach, the delay for each time step is calculated as the number of vehicles occupying each cell minus the number of vehicles that exit the cell. It is assumed that the traffic volume is fully known in each cell. A genetic algorithm was used to solve the network model on a simple single-direction network.

Li and Prevedouros (2004) used a hybrid optimization and rule-based strategy to control an isolated oversaturated intersection. The queue length of each lane from each approach must be measured or known. Their objective was to maximize the throughput and control the queue length. The calculation was conducted and the signal updated at the end of the active phase. Each phase green time is allocated as the time needed to disperse the maximum lane queue length subject to the maximum phase length; also the total number of vehicles served was calculated for each phase and then divided by the phase green time to get the throughput. MOVA (1986) used a similar detection method for isolated intersection signal control.

Park, et al. (1999) formulated the traffic network as an optimization problem. The cycle length, offset, green split and phase sequence were the decision variables, and the objective was throughput maximization, average delay minimization and modified average delay minimization with a penalty function. They developed a genetic algorithm to solve the network optimization problem numerically, in the absence of a closed-form solution for delay.

Researchers have tried to improve the utilization of the capacity of an intersection to relieve the congestion problem. But in oversaturated conditions, even with efficient use of the intersection capacity, vehicles still accumulate in every subsequent cycle and the residual queue on these approaches will finally reach the upstream intersection.

There has been some research directed towards the minimization of the spreading of network congestion from an oversaturated intersection. This research has sought to identify and evaluate corrective methods for preventing upstream queue effects, and to manage the effect of the departure flow on downstream signals. The concept of network queue management was formalized in the work by Lieberman et al. (1992, 2000) where the researchers first identified internal traffic metering to maintain stable queues in the congested network, and then developed a mixed-integer linear program to optimize the offsets. A nonlinear quadratic model was used to find the optimal phase durations. Arterial simulation was conducted with the WATSim micro simulation model to demonstrate the effectiveness of this approach compared to traditional signal timing models such as SYNCHRO, PASSER and TRANSYT that do not directly consider oversaturated movements and network congestion.

More recently, manufacturers have responded to suggestions from practitioners to provide features to address congested conditions, where traditional actuated fails. These methods are described in Chapter II, but one feature has been evaluated in recent research and should be mentioned here. Yun, et al. (2007) evaluated the dynamic maximum green in actuated controllers using hardware-in-the-loop simulation, finding that it outperforms commonly implemented alternatives. Essentially, this approach allows the maximum green time(s) of phases that serve oversaturated approaches to increase by some step size each cycle, up to some maximum total increase.

Among all these approaches, none except MOVA and the dynamic maximums are implemented outside the research context and few of those have moved beyond laboratory experimentation. None except the research into adaptive maximum green were developed in the context of current traffic control hardware, and most would require implementation at the system rather than intersection level. Many require detection capabilities that do not exist in practice, such as the ability to measure queue length.

Below is an annotated bibliography of papers referred to above that have addressed the issues of oversaturated intersections or congested networks, followed by a general list of references relevant to control methods at saturated intersections. The final section of this chapter includes a special discussion of NCHRP 3-66, which has reconsidered intersection control from the perspective of the latest detection capabilities.

Annotated Bibliography

Foundational Research

Gazis, D. C. and R.B. Potts, "The Oversaturated Intersection", *Proceedings of the Second International Symposium on the Theory of Traffic Flow*, London, England, 1963

Gazis, D. C., "Optimal Control of a System of Oversaturated Intersections", *Operations Research* Vol. 12, pp. 815-491, 1964.

The oversaturated isolated intersection control was first studied in this paper. The intersection under study consists of two phases with no turns, and assumed that the demand rate Q(t) is known and it follows a linear function. The method tries to minimize the total delay over the oversaturated period T. The resulting solution divided control into two stages over the time T, an optimal switchover time from stage one to stage two is found. In the first stage phase 1 is assigned maximum green time phase 2 is assigned minimum green time, while in the second stage the phase 1 is assigned minimum green time and phase 2 is assigned maximum green time. In Gazis (1964), this approach is extended to a system of oversaturated intersections. No simulation was conducted to study the performance of such control strategies.

Chang, Tang-Hsien and Jen-Ting Lin, "Optimal Signal Timing for an Oversaturated Intersection", *Transportation Research Part B*, Vol. 34, pp.471-491, 2000.

This paper reformulated the problem in Gazis (1963) using a discrete state space method and developed an algorithm to solve the model. In the discrete model the switchover time occurred exactly at the end of a cycle.

Optimization Model Research

Lo, Hong K., M.ASCE, and Andy H.F. Chow, "Control Strategies for Oversaturated Traffic", *Journal of Transportation Engineering*. August 2004.

This paper applied the Cell Transmission Model (CTM) to the control of oversaturated traffic network. The link between intersections is divided into cells with length of the travel distance at free flow speed, the number of vehicles in cell in each time step is monitored. Given this information the model optimize the signal of whole network. The genetic algorithm is used to find a near optimal solution; it is capable of generating variable green splits with no cycle plans. Simulation is conducted using a real traffic network consisting three intersections in Hong Kong and the result is compared with existing timing plan.

Hong Li and Panos D. Prevedouros. "Traffic Adaptive Control for Oversaturated Isolated Intersections: Model Development and Simulation Testing",. *Journal of Transportation Engineering*. Sep 2004

Traffic adaptive control for oversaturated intersections called TACOS is described and simulated. In this approach, there is no fixed cycle length, and phase sequence is dynamically assigned. The queue length on each lane must be estimated, and the phase green duration is calculated as the time needed to discharge the longest queue associated with that phase plus the time needed to service vehicles that joining the queue during the queue dissipation. The phase duration is also subject to min and max constraints. The method estimates the total number of vehicles to be serviced for each candidate phase, given phase green duration and total number of vehicles that need to be serviced for a phase. The algorithm calculates the throughput near the end of each phase, and chooses the next phase that provides the maximum throughput.

The algorithm is simulated in an environment called ICS which emulates NETSIM and INTEGRATION and was designed to be able to simulate the intersection operation under pretimed, actuated and TACOS control. The simulation compared the result of these three control strategies, claimed that the TACOS improve the performance significantly.

It is not known yet how TACOS performance compared to the other adaptive control strategies. But compared to some other adaptive control strategies, TACOS does not require the prediction algorithms which may introduce extra errors, even though prediction allows the controller to respond to the traffic proactively and therefore minimize the number of stops and delay.

TACOS did not address the solution for coordination if there is a need to be extended to arterial or network environment. The algorithm was designed only for congested conditions, and was not evaluated for sub-capacity conditions.

Yun, Ilsoo Matthew Best, and Byungkyu "Brian" Park. "Evaluation of the Adaptive Maximum Feature in the EPAC300 Actuated Traffic Controller Using hardware-in-the-Loop Simulation", TRB poster session, 2007

When operating under heavy traffic volume from all approaches, an actuated controller tends to function as a fixed time controller with maximum green time on saturated approaches. With adaptive max, the max green is incrementally changed according to the gap out information. Contiguous maxing out results in increasing max times, and contiguous gapping out results in decreasing max times. By using this method, more green time could be allocated to the saturated approaches, thus reducing the congestion. This paper evaluated adaptive max scenarios and identified research problems for dynamic max, such as the appropriate step size and upper and lower bounds of the resulting maximum green value. Simulation was conducted using hardware in the loop with VISSIM. The adaptive maximum was compared to two other alternatives: an arbitrarily large maximum green and the maximum green suggested by SYNCHRO. The result showed the adaptive maximum resulted in improved performance.

Network Metering

Edward B. Lieberman and Carroll J. Messer. *NCHRP 3-38(4) Final Report: Internal Metering Policy for Oversaturated Networks.* TRB, National Research Council, Washington DC, 1992

Edward B. Lieberman, Jinil Chang, and Elena Shenk Prassas. "Formulation of Real-Time Control Policy for Oversaturated Arterials", *Transportation Research Record 1727,* TRB, National Research council, Washington,D.C.,2000 pp.77-88

In this work, researchers formulated a control policy they called IMPOST, which implements an approach for metering traffic into congested intersections. The objectives are maximization of system throughput, full use of storage capacity to control the queue growth, and equitable service to cross street traffic and left turn traffic. A mixed-integer linear program (MILP) is used to find optimal signal offsets and queue length for each approach. A separate quadratic programming model is used to adjust the arterial green phase durations of each signal cycle so that the actual arterial queue lengths on each saturated approach will continually and closely approximate the optimal queue length computed by the MILP formulation. The simulation compared four different signal timing tools, including IMPOST, PASSER, TRANSYT, and SYNCHRO. The result showed that the IMPOST policy yielded improved network travel speed and delay under congested conditions.

Genetic Algorithms

Byungkyu "Brian" Park, Carroll J. Messer, and Thomas Urbanik . "Traffic Signal Optimization Program for Oversaturated Conditions - Genetic Algorithm Approach", *Transportation Research Record 1683,* TRB, National Research council, Washington, D.C., 1999 pp. 133-141

Byungkyu "Brian" Park, Carroll J. Messer, and Thomas Urbanik   "Enhanced Genetic Algorithm for Signal Timing Optimization of Oversaturated Intersections", *Transportation Research Record 1727,* TRB, National Research council, Washington,D.C., 2000 pp. 32-41

The program tries to optimize the cycle length, phase split, phase sequence, and offset simultaneously by using a genetic algorithm. The objective function value is evaluated using a mesoscopic simulator instead of an explicit deterministic objective function. The program considers the optimization of the four parameters of each intersection in a network. The paper compared the genetic algorithm approach with TRANSYT-7F optimization evaluated using CORSIM. The research showed that for low, medium and high flow conditions, the genetic algorithm performed best in low and high conditions, and performed as well as TRANSYT-7F in medium flow conditions.

In the second paper, the objective function is modified to prevent the low demand movements from being unrealistically delayed.

MCH1542 B Installation Guide for MOVA
MOVA was developed by the Transport Research Laboratory (now the Transport and Road Research Laboratory) to specifically address isolated intersection control. The MOVA controller, though not compatible with U.S.-style signal controllers, has been deployed in the field in England and other countries throughout the world.

The controller uses different strategies for undersaturated and oversaturated conditions. In the undersaturated condition the goal is minimization of vehicle delay and vehicle stops for the whole intersection. In the oversaturated condition the objective is to maximize the capacity on the congested approaches.

The controller uses lane-based detection. Two detectors are needed for each lane within a specified travel time from the detectors to the stop line at cruise speed. The first one is 8 seconds and second one is 3.5 seconds. The controller primary has a preset sequence of stages, depending on demand the stage maybe skipped.

MOVA's logic checks if any approaches are saturated, and if so maximum green time will be allocated to that approach. MOVA includes logic to determine the green duration of other non-saturated approaches in such a way that the green time is efficiently used.

Red Clearances
Two studies have been conducted that suggest that there is no longer term net benefit to increasing red clearances beyond the normal practice of providing clearance time across the intersection. Increases in red clearances were noted later in this project to have a disproportionate effect on congested movements, as explained in Chapter II. These studies include:

Roper, Brian A., Jon D. Fricker, C. Sinha Kumares, Robert E. Montogomery. The effects of the All-Red Clearance Interval on Intersection Accident Rates in Indiana. Report No. FHWA/IN/JHRP-90/7, Indiana Department of Transportation.
http://ntlsearch.bts.gov/tris/record/tris/00623613.html

Reginald R. Souleyrette, et. al, Effectiveness of All-Red Clearance Interval on Intersection Crashes, Report No. MN/RC-2004-26, Minnesota Department of Transportation, May 2004. www.lrrb.org/PDF/200426.pdf

References

Abu-Lebdeh, Ghassan and Rahim F. Benekohal. "Genetic Algorithm for Traffic Signal Control and Queue Management of Oversaturated Two-Way Arterials" In *Transportation Research Record: Journal of the Transportation Research Board No.* 1727, TRB, National Research Council, Washington, DC, 2000, pp.61-67

Abu-Lebdeh, Ghassan and Rahim F. Benekohal. "Signal Coordination and Arterial Capacity in Oversaturated Conditions" In *Transportation Research Record: Journal of the Transportation Research Board No.* 1727, TRB, National Research Council, Washington, DC, 2000, pp.68-76

Abu-Lebdeh, Ghassan and Rahim F. Benekohal. "Design and evaluation of dynamic traffic management strategies for congested conditions" *Transportation Research Part A* 37 2003, pp.109-127

Chang, Tang-Hsien and Jen-Ting Lin. "Optimal signal timing for an oversaturated intersection" *Transportation Research Part B 34,* 2000, pp. 471-491

Choi, B.K. *Adaptive Signal Control for Oversaturated Arterials*, Ph.D. Dissertation. Polytechnic University, Brooklyn, NY, 1997

Daganzo, C.F. " The cell transmission model: A dynamic representation of highway traffic consistent with the hydrodynamic theory." *Transportation research part B 28B(4),* 1995, pp. 269-287

Gazis, D. C. and R.B. Potts. "The Oversaturated Intersection" *Proceedings from the Second International Symposium on the Theory of Traffic Flow*, London, England, 1963, pp. 222-237.

Gazis, D. C. "Optimal Control of a System of Oversaturated Intersections" *Operations Research 12,* 1964, pp. 815-491.

Li, Hong and Panos D. Prevedouros. "Traffic Adaptive Control for Oversaturated Isolated Intersections: Model Development and Simulation Testing" *Journal of Transportation Engineering.* September 2004, pp. 594-601

Lieberman, Edward B., A.K. Rathi, G.F.King, and S.I. Schwartz. "Congestion-Based Control Scheme for Closedly Spaced, High Traffic Density Networks" In *Transportation Research Record: Journal of the Transportation Research Board No.1057,* TRB, National Research Council, Washington, DC, 1986, pp.49-57

Lieberman, Edward B. and Carroll Messer. *Internal Metering Policy for Oversaturated Networks—Final Report.* Project 3-39(4), National Cooperative Highway Research Program, TRB, National Research Council, Washington, DC, 1992,

Lieberman, Edward B., Jinil Chang, and Elena Shenk Prassas, "Formulation of Real-Time Control Policy for Oversaturated Arterials". In *Transportation Research Record: Journal of the Transportation Research Board No.1727*, TRB, National Research Council, Washington, D.C., 2000, pp.77-88

Lo, Hong K. and Andy H. F. Chow. "Control Strategies for Oversaturated Traffic", *Journal of Transportation Engineering*. American Society of Civil Engineers, August 2004, pp. 466-478.

Michalopoulos, P.G., Stephanopolos, G., "Optimal control of oversaturated intersections theoretical and practical considerations," *Traffic Engineering & Control*. May 1978, pp. 216-221

MCH1542 B Installation and Application Guide for MOVA

Park, Byungkyu, Carroll J. Messer, and Thomas Urbanik . "Enhanced Genetic Algorithm for Signal Timing Optimization of Oversaturated Intersections." *Transportation Research Record No.* 1727, TRB, National Research Council, Washington, DC, 2000, pp.32-41

Rouphail, Nagui, Andrzej Tarko and Jing Li, "Traffic Flow at Signalized Intersections" *Traffic Flow Theory Monograph,* ed. Nathan Gartner and Carroll Messer. First draft, printed for Federal Highway Administration by Oak Ridge National Laboratory, 1995.

Yun, Ilsoo, Matthew Best, and Byungkyu Park. "Evaluation of the Adaptive Maximum Feature in the EPAC300 Actuated Traffic Controller Using hardware-in-the-Loop Simulation", TRB poster session 2007

Overview of NCHRP 3-66

The NCHRP 3-66 project entitled *Traffic Signal State Transition Logic Using Enhanced Sensor Information* focused on three related issues in traffic signal control. First, an architecture for traffic signal control logic was developed to support the discussion and development of logic. Next, a model of the core traffic signal controller logic was developed and used to show how single and multiple priority requests can be served and how the different phase intervals, minimum, pedestrian, etc. relate to the signal timing logic. The role and use of traffic signal detectors was investigated and improved detection information was defined and linked to the controller core logic. Finally, the Highway-Rail interface was studied and recommendations for improved efficiency and safety were made.

The traffic signal controller architecture was developed to help the research team understand the different roles and responsibilities of the components of signal control logic. **Figure I-1** shows the key functional packages within the controller architecture. In the center, the Core Logic represents the standard ring-phase-barrier model of a controller. Each phase is constructed from intervals for vehicle and pedestrian indications. The Strategic Control package represents the logic that would is responsible for providing strategic control such as coordination, preemption, and priority. The Priority Request Generator represents logic from buses, trains, trucks, snowplows, or other vehicles that might request special service from the intersection. The Tactical Control package utilizes Sensors (detectors – loop, video, or other technologies) to determine when to call phases for service and when to call for extension. Control of oversaturated isolated intersections is primarily an issue in the Tactical Control and Core Logic components in this architecture. Control of networks of oversaturated intersections would include the Strategic Control Package.

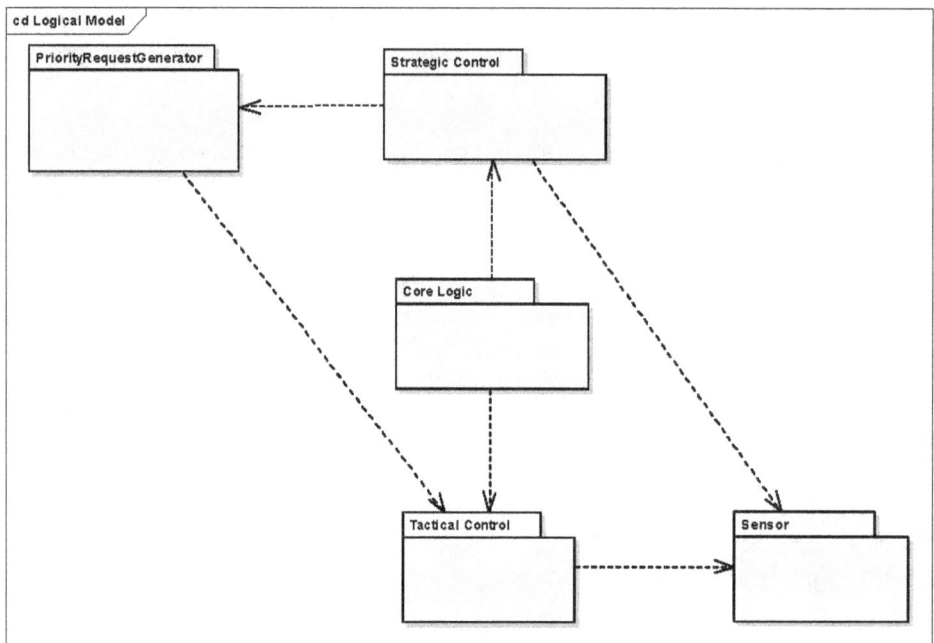

Figure I-1. Traffic Signal Controller Architecture

A significant effort addressed in the 3-66 research was the Tactical Control logic in terms of how detectors are used to call and extend phases. Several Tactical Control enhancements were investigated including the use of lane-by-lane logic to extend and gap out phases and flow based logic that could request a phase be terminated if the there were vehicles calling a phase, but the vehicles were not flowing due to some obstruction – such as downstream congestion or a blocked lane. These improvements in the Tactical Control logic improve the utilization of green time at the intersection significantly and will be important to the both short term and long term strategies for control of oversaturated intersections.

The 3-66 contributions to the Highway-Rail interface investigated Sensor issues related to track circuits to determine how far in advance a Request for Service could reliably be made to allow safer and smoother preparation for the required Simultaneous Preemption Call. The Core Logic model was used as the foundation for advanced preemption service using concepts from project scheduling to determine which phase intervals could be served, such as pedestrian walk and flashing don't walk and phase extension intervals, and still satisfy the Simultaneous Preemption Call.

The 3-66 research provides a logic framework for consideration of strategies and algorithms for control of oversaturated intersection. The controller architecture helps allocate functionality to the detection and tactical decision making components and to the phase control logic. The Core Logic model provides a decision framework for servicing phases intelligently under oversaturated conditions. The Tactical Control package provides the necessary interface and relationship for future extensions to arterials and networks.

Key findings related to oversaturated control from the 3-66 research primarily include tactical control and sensors; however, the structure of the overall architecture was designed to support analysis of all levels of control and are directly applicable to oversaturated control.

The role of sensors in traffic control is to provide information about the state of traffic demand for service by the shared capacity that is controlled by a traffic signal. Today, most sensors provide tactical input as pulses or present as vehicles cross a defined point in one or more lanes. For example, a stop bar detector can cover multiple lanes on an approach to an intersection and provides a CALL signal to the tactical control level of the controller. The CALL may be active as long as there are vehicles on the approach or may be LOCKED (during the yellow change or red clearance intervals) and cleared when any green (service) interval serves. An upstream passage detector may send a pulse signal to the tactical control unit that may be used to extend a phase if it is during the extension interval or to CALL a phase during the red interval.

The NCHRP 3-66 research addressed two improved roles for sensors including lane-by-lane detection and flow based measurement at the tactical level and a role for sensors in the strategic level including the estimation of volume to capacity (v/c) ratios that are used in the development of signal timing. The idea behind lane-by-lane detection is to address the issues related to lane utilization where one, or more, lanes may be more heavily utilized rather than an equal utilization across all lanes. The lane-by-lane concept also addresses the efficiency of utilization of green time by allowing each lane to "gap out" instead of allowing a ping-pong effect of extension calls in each lane to fully extend the green phase to the maximum green time, yet actually only serving a fraction of the service capacity. **Figure I-2** (adopted from the NCHRP 3-66 Final Report) shows this effect. This concept is applicable more in the under saturated conditions than oversaturated conditions, but is relevant during periods leading to and from oversaturation.

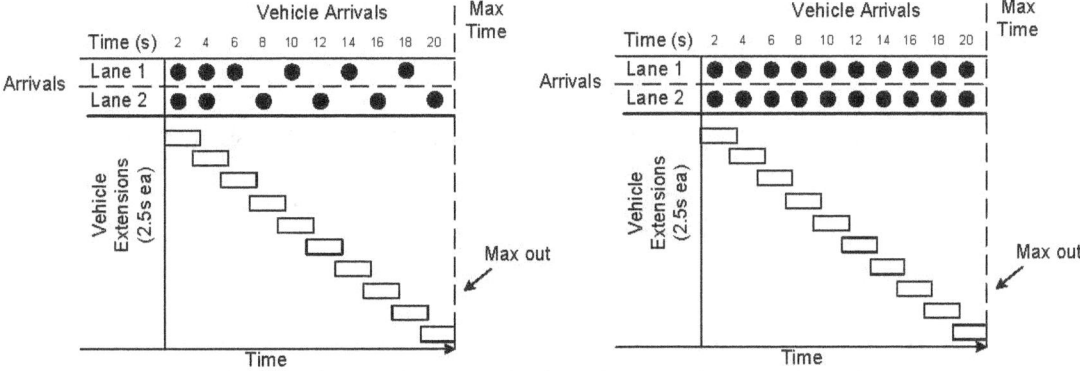

Figure I-2. Ping-pong effect of lane utilization in traditional detection.

Flow based sensors would provide information about the number of vehicles that cross the detector. The concept is based on the use of stop bar detectors to measure flow during the green service interval and the request to terminate green if there is presence but no flow. For example, assume that the downstream link is congested and a movement (e.g. left turn) cannot discharge any vehicles. Traditional control would allow the phase to extend to the maximum green time despite the fact that no vehicles are departing and the green time might be used for another movement. The NCHRP 3-66 research showed that in this condition, the green could be terminated and control could move onto other phases. This might reduce the overall cycle time, hence increasing the frequency of phase service while reducing the duration of service with the effect of increasing the true phase green utilization as measured by flow of vehicles during the green interval. Coordinated operations might constrain this effect by moving the additional green time to the coordinated phase(s) hence reducing the duration, but not increasing the frequency of service. This issue was not addressed by the NCHRP 3-66 project.

The role of sensors in strategic control studies in the NCHRP 3-66 research included the dynamic reallocation of phase split times (green times) based on the measurement of volume to capacity (v/c) ratios. This role addresses the fundamental issue of how traffic signal timings are developed. Currently, a traffic study is conducted where traffic counts (volume and turning counts) are collected on a network over a period of several days. Typically this is done manually, however there are some new tools for automated data collection that are used. These count represent point estimates of the demand for different movements that are used in signal timing software (such as Synchro, TRANSYT, PASSER, etc.) to develop the timings. These point estimates fail to capture the random variations that can occur over time, or on a daily basis and the changes that occur over longer time periods such as weeks, months, or years. The ability to measure and utilize flow data at the intersection could allow the strategic level to reallocate green time to address these variations. Again, this analysis did not address the oversaturated conditions and strategies that could be used to detect (such as phase failures) or address oversaturated phases.

In addition to the NCHRP 3-66 research, Abbas, et.al. looked at the use of system detector data (defined as lane-by-lane volume and occupancy data collected over a one minute of one cycle period at upstream detectors) to evaluate the effectiveness of offsets between intersections. Essentially if the volume was high and the occupancy was low then the offset must be allowing the vehicles to progress through the intersection. If the volume was low and the occupancy is high, then the offset may not be allowing the progression of vehicles and contributing to potential oversaturated conditions.

The core logic model developed in the NCHRP 3-66 research provides an analytical structure for understanding and analyzing the capability of the traffic signal controller. Head, et. al. (2007) represent this model as a precedence graph that is similar to the activity diagram used in project management and critical path analysis. Figure I-3 shows an example of the precedence graph model.

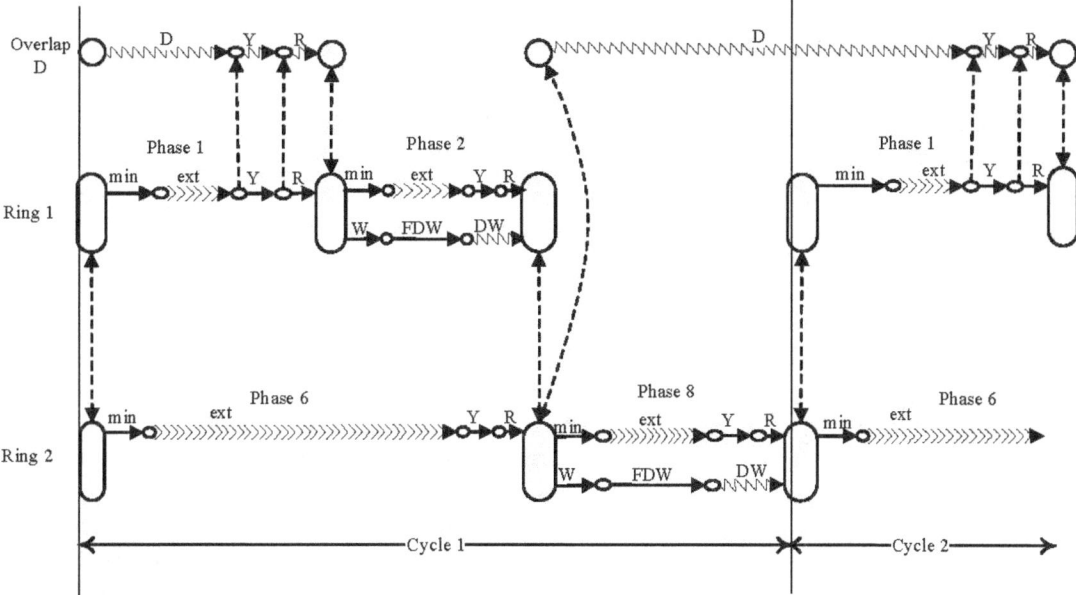

Figure I-3. Precedence graph model for T-intersection (from Head, et. al. 2007).

This model shows how intervals, such as pedestrian intervals, can constrain the early termination of a phase – as might be requested by the tactical level when there is low or no flow across and stop bar detector. It also helps to show how one phase, or phase group such as phases 1 and 2 in ring 1, may constrain the termination of other phases such as phases 3 and 4 in ring 2. As various tactical and strategic controls are evaluated the interaction and effect of the structural issues are captured in the model. The model is descriptive of controller behavior and not prescriptive of how the controller should operate in oversaturated or undersaturated conditions.

References for NCHRP 3-66 Discussion

Abbas, M., Bullock, D., Head, L, "Real-Time Offset Transitioning Algorithm for Coordinating Traffic Signals", *Transportation Research Record No. 1748,* TRB, National Research Council, Washington, DC, 2001.

Head, K.L., D.G. Gettman, D. Bullock, and T. Urbanik, "Modeling Traffic Signal Operations Using Precedence Graphs", to appear, *Transportation Research Record,* TRB, National Research Council, Washington, DC, 2007.

Head, K.L., D. Gettman, and Z. Wei, "A Decision Model For Priority Control of Traffic Signals", *Transportation Research Record No. 1978,* TRB, National Research Council, Washington, DC, 2006, , pp. 169-177.

Smaglik, E., Bullock, D., Sturdevant, .J, Urbanik, T., "Implementation of Lane-By-Lane Detection at Actuated Controlled Intersection", to appear, *Transportation Research Record,* TRB, National Research Council, Washington, DC, 2007.

II. The State of the Practice

Categorizing Congestion Based on Operational Objectives

Identifying strategies for mitigating the effects of saturated signal approaches requires an understanding of the objective of operational decisions. These objectives vary with conditions, though this variation is rarely articulated by practitioners or used in recommendations to practitioners. They are, however, implicit in the decisions made by practitioners.

Over the history of the practice, methods that have received wide use already distinguish between a range of objectives in signal control. For example, a traditional approach for calculating cycle length at isolated intersections, first taught at the Yale Bureau of Highway Traffic, determines the expected arrival platoon at a signal using the Poisson distribution. This method assumes independence between arriving cars and thus is only valid at low arrival flows. It identifies the arrivals during a cycle that will only be exceeded 5% of the time, and then provides green time sufficient for those arrivals. The objective of this function is to serve 95% of the arriving platoons, which stated another way is to minimize cycle failures. Cycle failures occur when the waiting queue is not fully served by the next green interval.

At some point, minimizing cycle failures becomes an unattainable objective. The Poisson method will drive up the cycle length, with the result that the longer red times will create longer queues to serve, which drives up the cycle even more. In the Poisson method, one iterates to find a cycle at which all the 95%-ile arrivals can be served, but those iterations fail when traffic demand reaches moderate amounts. And at those demand levels, the assumptions underlying the Poisson method are no longer valid, in that the cars are no longer behaving independently and the Poisson distribution no longer describes their arrivals.

During such conditions, most practitioners switch to a method that seeks to achieve smooth flow consistent with driver expectations. This objective may be explicitly represented by minimizing delay, stops, or some combination of the two. Delay and stop minimization is a different objective than minimizing cycle failures, because at some point serving a long departing queue causes more delay on the competing movements than it saves on the approach being served. Most traffic signal timing tools currently available seek to minimize some combination of stops and delay.

Some tools use degree of saturation on the approach as a surrogate for stops and delay. For example, the Highway Capacity Manual is based on the fundamental relationship of G/C*S, where G/C is the ratio of green time to cycle length and S is the saturation flow. This approach seeks to define the capacity of the approach, and signal timing based on this approach attempts to balance the saturation of each movement. The objective is therefore to minimize the difference in the degree of saturation for the various approaches at an intersection, and those variations are a negative performance measure within that approach.

PASSER II has as its main optimization objective signal progression along an arterial street (more on that below), but it seeks to balance the degree of saturation at each intersection to optimize splits. Both are closely related to Critical Lane Analysis, which first appeared in TRB Circular 212, *Interim Materials on Highway Capacity*. This approach explicitly *does not* minimize delay. For example, if an intersection with four approaches is served by a two-phase signal, and three of the approaches each serve 1000 cars while the fourth only serves 100 cars, balancing the degree of saturation will result in equal green times for the two intersecting roadways, assuming no left turns. But when the street carrying 1000 cars in one direction and 100 cars in the other direction is green, only 1100 cars are being served, while 2000 cars are being delayed. Minimizing delay would provide close to two-thirds of the cycle to the street with two 1000-car/hour approaches, and one third to the street carrying only 1100 cars. This approach might, however, cause congestion on the 1000-car/hour approach that is opposing the light movement.

Within coordinated systems, optimization approaches consistent with achieving smooth flow optimize to minimize delay and stops in the network or to optimize a specific smooth-flow objective, such as progression. Progression is the ability of cars to pass through a series of coordinated signals without being stopped by a red signal. All such opportunities form a band that is the portion of the cycle during which vehicles traveling at the desired speed can achieve progression through the intersections in question. As mentioned previously, PASSER II explicitly maximizes bandwidth when determining the coordination between intersections and the phase sequence at each intersection. The splits at each intersection are determined to provide balanced saturation, however.

Optimizations that seek to minimize delay and stops, maximize progression, or balance the degree of saturation, fail when an approach becomes congested. The reason is that neither delay nor saturation can be evaluated during congestion, and progression cannot be achieved through a standing queue. Saturation is limited by the capacity of the signal, and when the capacity is exceeded, saturation no longer characterizes demand. Delay equations are indefinable in congested conditions because the delay increases as long as the demand exceeds capacity. Delay equations require a stable relationship between vehicles arriving to be served and those departing, having been served. When the system is storing more and more cars because of a capacity constraint, delay will continue to increase for the subsequently arriving vehicles.

Thus, an important question is determining at what point a practitioner should change objectives from attempting to achieve smooth flow to strategies defined specifically for congested conditions, when smooth flow is impossible.

It should be noted that the boundary between shifting objectives is not fixed, and the operation can cross the boundary as a result of what practitioners do. Thus, one critical strategy apparent in the case studies is to avoid inducing congestion when smooth flow is still attainable.

This chapter discusses a series of case studies and a series of interviews with practitioners in the context of these objectives. These case studies will be evaluated on the basis of which objective was being attempted, and whether the results of those attempts were successful. Several of the case studies show the progression from smooth-flow strategies that failed to congestion strategies.

In addition to identifying strategies by their objectives, the cases studies and interviews in this chapter will be used to catalog a series of action points to reveal when a change in strategies might be critical. These action points are closely related to the definition of saturated conditions.

NCHRP 3-38(4) attempted to define saturated conditions as follows:

- Local congestion. This occurs when one or more approaches faces a cycle failure that does not result in damaging or excessive queues.
- Extended congestion. Cycle failures repeat such that queues extend damagingly through upstream intersections, causing the capacity of those intersections to be reduced.
- Regional congestion. This occurs when the queue from a critical intersection joins or influences the queues at upstream *critical* intersections. In other words, the queuing from one critical intersection imposes a cascading capacity constraint on other intersections which were at the saturation point even without the influence of the spreading queue.
- Intermittent congestion. This occurs as a natural result of stochastic traffic arrival. Even at light volumes timed using the Poisson method, one expects cycle failures 5% of the time. McShane and Roess, in *Traffic Engineering*, demonstrate this using a diagram similar to Figure II-1. They describe this as a brief period of congestion in an otherwise stable traffic stream. Distinguishing between intermittent and prolonged congestion will likely affect the selection of responses, and thus is an important distinction to be made.

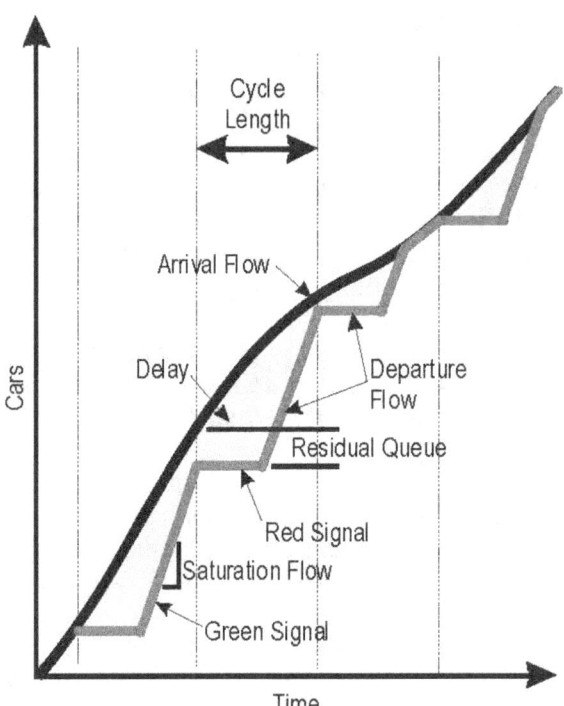

Figure II-1. Intermittent congestion in stable traffic stream

- Recursive (cyclical) congestion. This is congestion at an intersection with insufficient capacity that occurs predictably as a result of foreseeable demand patterns.

- Prolonged congestion. Congestion in the network is relentless and queue interactions create such inefficiencies that demand must fall well below the reduced capacity for extended periods to permit residual queues to clear.

These definitions served the purpose of that research but are not necessarily clearly related to actions that practitioners would take. A useful feature of these definitions is that it distinguishes between occasional cycle failures that will be present even in moderate traffic conditions and the sort of building residual queues that compel action by the practitioner. One problem is that the order in which they are presented do not provide a clear hierarchy of severity, nor do they clearly suggest to the practitioner to change objectives. A review of the NCHRP 3-38(4) work suggests that these categories were define mostly to link to specific traffic signal system control strategies, not all of which are currently relevant.

Also, the conditions described in NCHRP 3-38(4) are closely related to coordinated signal timing, while the purpose of this research is to explore intersection-based strategies.

Thus, these levels can be simplified, providing a hierarchy of conditions keyed to action levels:
- **Light traffic** Characterized by the expectation of minimized cycle failures. The objective in these conditions is to fully serve arrival queues. Cycle failures are expected to be infrequent at less than 5%.
- **Moderate traffic** Characterized by the expectation of "fair" operation. At these intersections, drivers expect the operation to be "fair", which usually means that the objective should be both obvious (or easily explained) and equitable. Progression in a network falls into this category, as does minimizing delay and stops, or balancing the degree of saturation on each approach. Some cycle failures are to be expected in meeting these other objectives and do not necessarily violate the expectations of motorists.
- **Heavy traffic** Characterized by frequent cycle failures, but with a residual queue that ebbs and flows without growing uncontrollably. The queue is the result of stochastic arrival flows where those flows exceed capacity up to about half the time.
- **Oversaturated operation** Characterized either by excessive residual queues that grow (seemingly at least) without control, or by queues that cause more widespread damage to the operation of a network. It is no longer possible to determine what percentage of the arrival flows can be served, because they usually do not reach the intersection during the green interval that might serve them.

These approaches can be characterized graphically, as shown in Figure II-2.

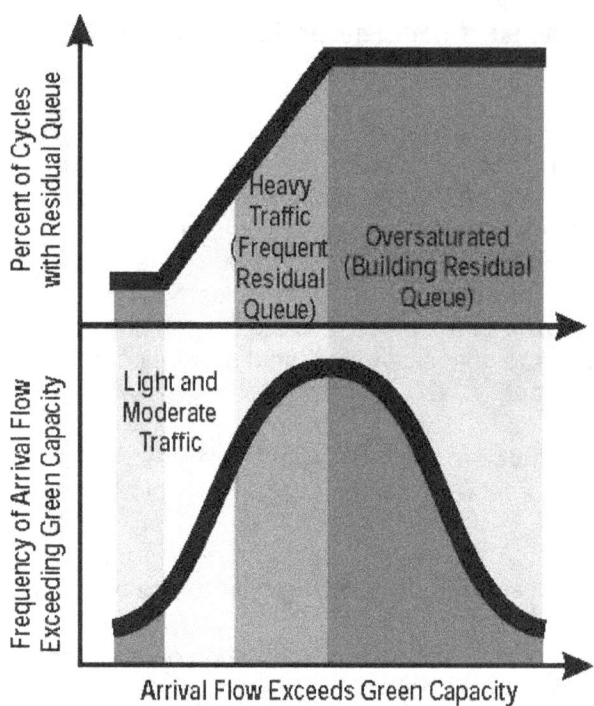

Figure II-2. Saturation and Residual Queues

Figure II-2 is not intended to represent actual statistics, but rather to illustrate that when average demand equals approach capacity, the percent of cycles with residual queuing will be 100%. In fact, it is likely that residual queuing is a certainty even when average demand is somewhat less than capacity. When average demand exceeds capacity, a growing residual queue is inevitable.

In addition to describing the conditions at which certain actions and objectives are compelled, the case studies and interviews reveal those actions. Of particular importance is determining whether the actions taken are intended to be curative or palliative. In the medical world, doctors treat patients with the objective of achieving a cure. When a cure is not possible, then they treat terminally ill patients with the objective of minimizing pain and suffering.

The initial expectation of the research is that actions taken to alleviate congestion are appropriate under heavy traffic conditions as described above, and are curative. The objective is to maximize throughput in order to minimize or eliminate residual queues. And if these actions fail, one expects that the objective will shift to minimizing the damage of queue formation to the extent possible, which is palliative. The analysis of the example applications and interview responses will be to determine the accuracy of these expectations.

Example Applications

Example 1. Square Lake Road at Telegraph Road, Oakland County, Michigan

In this case study, the wide median of the Michigan "boulevard" prevented normal opposing left turns. Left turners first turn right and then U-turn to complete their maneuver. Thus, left turners approach the traffic signal twice.

Before improvements were made, the operation started with westbound green and a green for the signalized U-turn. This provided uninterrupted flow for left turners to first turn right and then make their U-turn. After a period of time, the U-turners were stopped to allow the southbound through movement to proceed up to the intersection. That southbound movement then received a green signal at the intersection.

Finally, the southbound movement was stopped at the U-turn to allow remaining U-turners to make their U-turn and still be able to go through the intersection with a southbound green signal.

In terms of its operational effect, this sequence first allowed left turners to turn right and then U-turn to then store in the southbound intersection approach. The southbound movement at the U-turn then turned green to serve southbound through cars. This condition prevailed for a while until the southbound movement at the intersection received a green. Thus, the operation was intended to provide southbound progression primarily, and depended on relatively small volumes making the right-turn-U-turn sequence of maneuvers. The problem was that the volume of U-turners increased to the point that it filled up the southbound approach at the intersection, which caused the queue to block the progression for southbound through cars. The southbound approach became congested, and the operation favored those making the left-turn maneuver at the expense of the southbound maneuver.

The objective of the before condition was to provide smooth flow for the southbound through movement, but that objective was thwarted by congestion from the oversaturated southbound approach and the queue formation from U-turners.

To minimize the problem, staff at the Road Commission for Oakland County altered the sequence. The altered sequence started by providing an open southbound green at both the U-turn and the intersection. This prevented the queue from the U-turn from blocking southbound through cars. After the southbound movement was served, the sequence served the U-turn.

The objective of the altered operation is to manage queue spillback to prevent the queue from a relatively unimportant movement from blocking a relatively important movement.

The sequence of movements that provided the most effective queue management is shown in Figure II-3.

Figure II-3. Signal Sequence for Queue Management, Square Lake Road and Telegraph Road.

Example 2. Sunrise Boulevard, Rancho Cordova, California

This congested corridor extends a little less than three miles through 11 signalized intersections from White Rock Road to Gold Country Boulevard. Traffic volumes on Sunrise increase from 1500 vph at the south end to 5000 vph at the north end in the peak direction. Sunrise Boulevard carries 80,000 vehicles per day on three basic lanes in each direction.

In the PM peak, demand exceeds capacity at the three northernmost intersections. Queue formation extends back from the first of these intersection upstream over the length of the section.

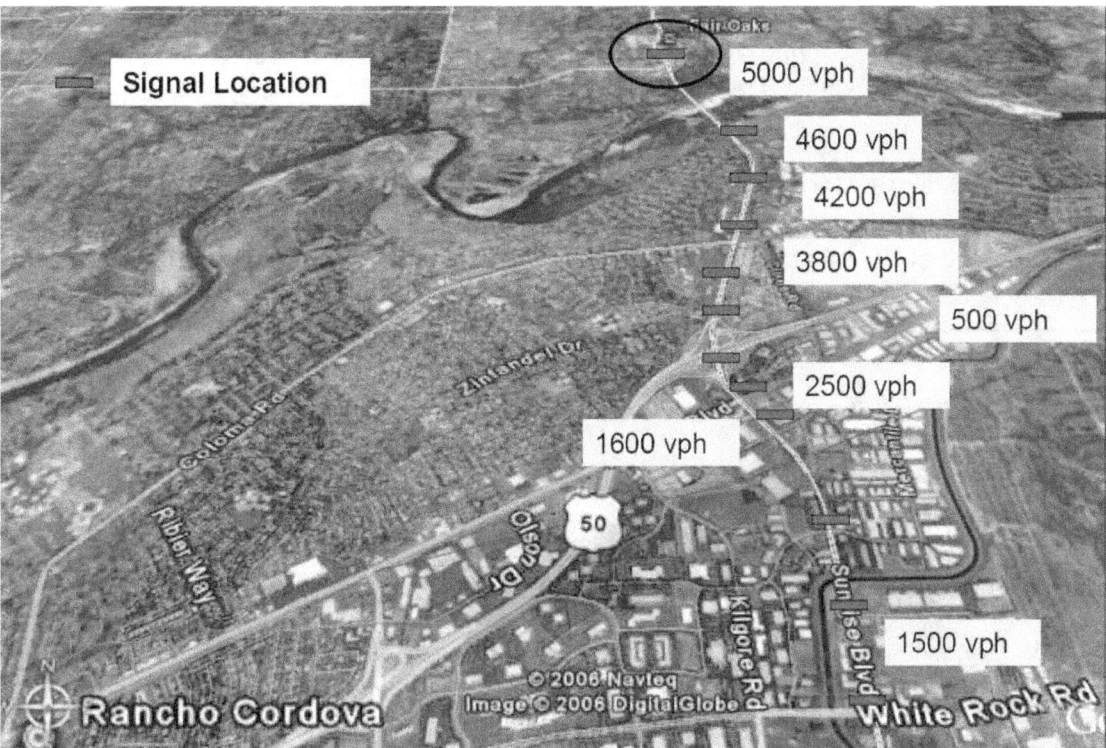

Figure II-4. Sunrise Boulevard Corridor.

The agency has considered widening, but is constrained by limited right of way.

Alternatives considered include:
- Optimized timing for all phases based on balancing saturation.
- Optimized offsets for improved coordination
- Shortened cycle length to try to minimize queue formation
- Longer cycle lengths to improve throughput, if possible.
- Left turn phase sequence
- Left turn protected-permissive operation
- Limited widening for northbound, despite lack of right of way
- Flush strategy of adding green time on Sunrise only

All but the final approach are curative, seeking to maximize throughput with the hope of minimizing or eliminating congestion.

The flush strategy is currently being used. It uses vehicle preemption to force the signal to pause while serving northbound Sunrise traffic. The flush is implemented manually. Nothing is done to alleviate the resulting congestion on side streets.

The desired strategy currently under consideration is an adaptive flush strategy implemented by time of day. The strategy would be defined by thresholds, possibly queue formation. Minor objectives would be to keep northbound turn bays clear, and to alleviate congestion on some major side streets. The objective is to focus queue formation on the portions of the roadway with the most favorable storage.

Both the manual and contemplated automatic flush strategies are palliative. They seek to move the damaging queues to the least damaging location. There is no expectation of alleviating residual queues, nor is there any expectation of maximizing throughput overall.

Example 3. North Carolina State Highway 54 and I-40, Durham, NC

The location in question is a network of three signals, two at the ramps of I-40 and NC-54, and one at NC-54 and Farrington Road, just west of the southbound ramp. See Figure II-5.

Figure II-5. NC 54 at I-40, North Carolina

The principle problem is a severe capacity constraint at the NC-54/Farrington intersection, which backs the queue back into the ramp intersections and onto I-40.

The existing operation attempted to provide progression at the start of green using a 153-second cycle length. Excessive queues formed westbound that backed up around the loop ramp to the northbound I-40 mainline. Both directions of Farrington Road also suffered residual queuing. Clearly, the objective to provide progression was not appropriate for the oversaturated conditions.

The first attempted corrective measure employed offsets intended to create storage for side-street traffic. This approach involved simultaneous reds on NC-54 so that the side streets would have access to the storage area. A key feature of this alternative was the use of shorter greens and cycle lengths to minimize turn-lane overflow. The green time was set to roughly empty the turn bays, and the red time set to roughly fill the turn bays. The cycle used was 110 seconds. This approach somewhat alleviated queue formation on Farrington, but the queue still backed out onto I-40, and the westbound left turn at Farrington still spilled out of its bay.

The first attempted approach was intended to avoid queue overflow in an attempt to maximize throughput. The hope (unrealized) was a cure for excessive queuing by maximizing throughput. A minor objective was to alleviate excessive delay on Farrington Road.

The second attempted corrective measure sought to assign capacity based on safety needs rather than delay. The alternative lagged the westbound left turn to reduce queue spillback and increased the cycle length to improve capacity. The cycle used was 150 seconds. The simultaneous main-street reds were retained, but the capacity on the Farrington approach was reduced.

The result of the second attempted approach was excessive queuing onto I-40 and excessive queues on Farrington Road.

This approach was still based on maximizing throughput to alleviate or minimize residual queuing. The objective was not met.

The third attempted strategy finally employed a pure queue management objective. The cycle was increased to permit a higher green percentage on westbound NC-54. Splits for movements that had safe storage were reduced. The controller was programmed with holds and recalls to prevent actuation from restoring green time to movements of less importance to the objective of safe queue storage. The cycle used was 200 seconds. The controller was programmed so that plan changes occurred on even multiples of the cycle length, with the same offset used in all plans. This technique eliminated plan transitions at critical intersections.

The result of the third solution was excessive queuing on Farrington Road, but lessening of the queue on westbound NC-54 such that the queue did not spill out into I-40 mainline.

None of the solutions provided acceptable operation using any objective other than queue management.

Example 4. Bandera Road and Guilbeau Road, San Antonio, Texas

Bandera Road is one of the major arteries serving northwest San Antonio west of IH-10. At the time of this example (early 1990's), the traffic signals along Bandera were coordinated for the first time. The problem was a very heavy outbound left turn movement, as depicted in Figure II-6. Widening to allow a two-lane left turn was not an option at the time, though doing so would have solved the problem. This case is typical of many, in that improvements to the lane utilization can often double the capacity of congested movements, while finding more green time is usually a marginal improvement.

Bandera Road

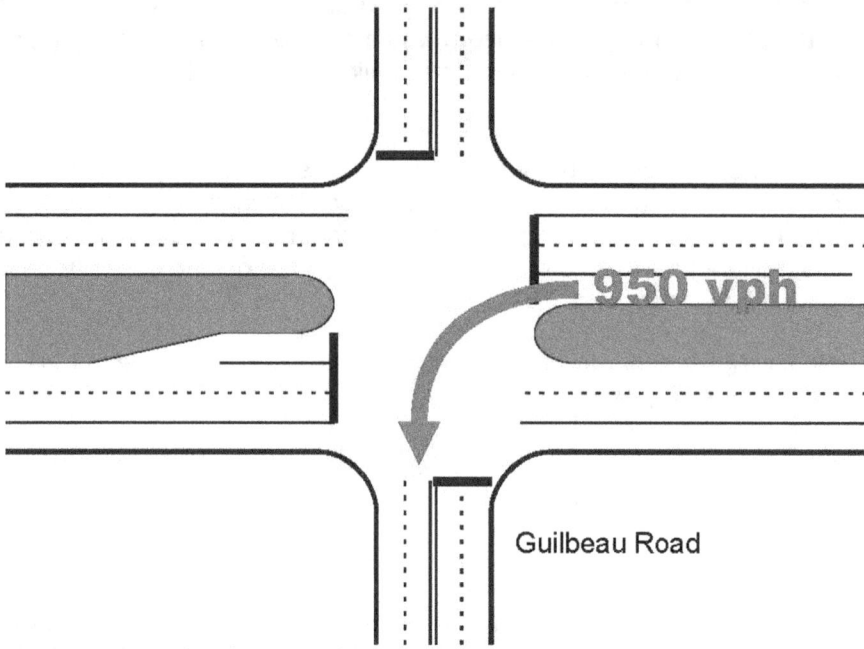

Figure II-6. Bandera Road and Guilbeau Road, San Antonio

The existing traffic signal phasing was conventional split phasing, given that most traffic on the side streets turned left to enter the main highway. The time required to serve the side-street movements exclusively consumed too large a share of the cycle, and the result was severe congestion on the outbound left turn and a growing residual queue that exceed a mile in length. When the queue spilled out of the left turn lane, it blocked a through lane, which caused congestion on the through movement.

The left turn volumes on the side street were light, and the first operational improvement was to eliminate the split phasing and require side-street left turners to yield. This resulted in a high level of public complaint, which compelled a different solution.

The next solution used phase reservice, which is a technique that appears over and over in discussions with expert practitioners, as described in the next section. Serving phases twice in the cycle may be used to reduce the number of times a minor movement is served, or increase the number of times a major movement is served, relative to the coordination. In this case, the intersection was operated at twice the normal coordination cycle of the system, and the main-street movements were served twice in the cycle. Thus, the effective cycle of the main street movements was consistent with system timing, but the percentage of that effective cycle consumed by the side street was significantly reduced.

Figure II-7 shows the final operation selected for this intersection.

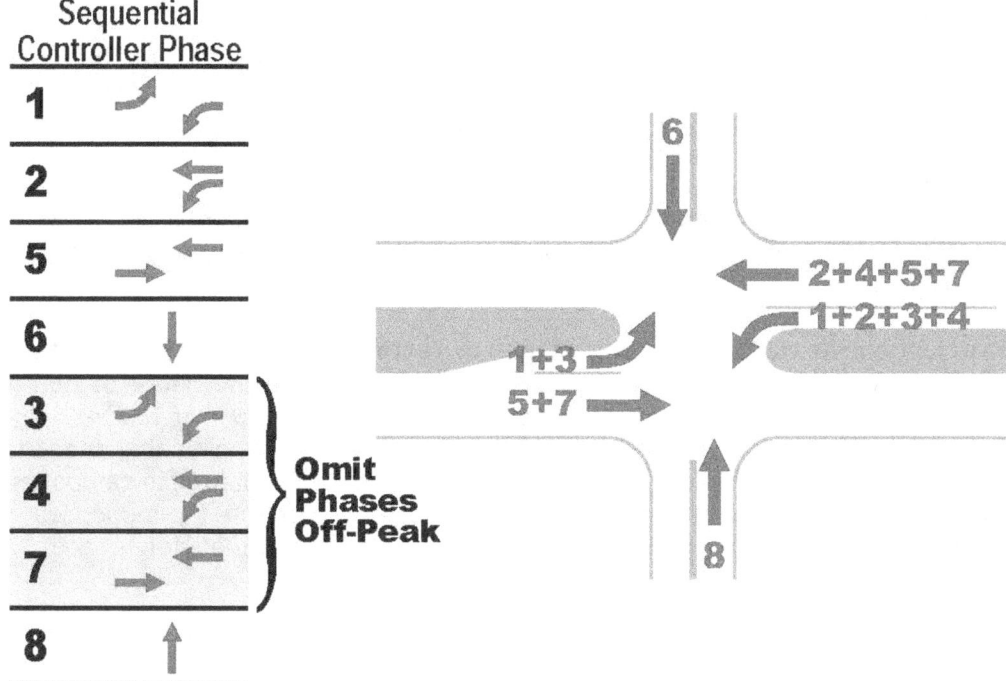

Figure II-7. Serving Major Phases More Than Once.

It should be noted that since this operation was implemented, the intersection was widened to provide a two-lane left turn. This operation, however, improved throughput until geometric improvements could be made.

State-of-the-Practice Interviews

Interviews were conducted with a variety of recognized experts. These experts are listed alphabetically below:

- Gerard de Camp, independent consultant, Las Vegas, Nevada and various locations in Texas
- Steven Click, Tennessee Tech University (formerly with North Carolina DOT), Cookeville, Tennessee
- Woody Hood, Maryland State Highway Administration, Annapolis, Maryland
- Eric Nelson, Harris County Department of Transportation, Houston, Texas
- Gary Pietrowicz, Road Commission of Oakland County, Troy, Michigan
- Ziad Sabra, Sabra-Wang and Associates
- Bill Shao, Los Angeles Department of Transportation, Los Angeles, California

The results reported below also include insights gained by the author while serving with the City of San Antonio in Texas.

A rough outline of questions was used as a means of promoting discussion. These are shown below:

1. Any potential case study or field test site information.

2. What conditions compel action?. Citizen complaint? Unmanageable queue formation? Queue spillback? Time in queue? Queue length? What is your definition of saturation?
3. What are you trying to accomplish with various action? What did you try? Why did you try it? How did you know if it did or didn't work?
4. What strategies and techniques do you implement?
5. What controller features do you employ? Do you go beyond conventional actuation?
6. What operational control strategies would you like to implement that are not supported by current controllers, or that require external manipulation of the controller?

In the actual interviews, the discussion ranged away from the questions significantly. This was allowed, because the objective was not to prejudge the knowledge of the respondents with leading questions, but to discover that knowledge in its native organization. Therefore, the questions provided a rough outline of points that needed to be included in the discussion. It was not productive, however, to attempt to categorize the interview responses according to the above questions. Therefore the responses summarized below are organized by themes that emerged during the interviews.

Definition of Saturation

The definitions offered by the interviewees varied in terms of specificity. These are some samples:

- Cannot serve all cars in sequential phases, after all attempts at balancing green times have been attempted.
- Exceeding maximum throughput.
- Damaging queue formation, particularly when the damage is a safety concern (e.g. a queue backing onto freeway main lanes) or risks gridlock by backing through an upstream intersection.
- Significant percentage of motorists needing more than one cycle to complete their maneuver.
- Growing residual queues on at least one movement after green times had been adjusted as much as possible.

All of these are related. In all cases, the observed phenomenon is growing residual queues. In these cases, delay cannot be evaluated. All define the saturated condition after making as many improvements as possible to green splits at the intersection. Most noted that green splits may be dictated by constraints unrelated to congestion on the approach.

With this definition, the first strategy was first to make as many conventional green-split improvements as possible before declaring the intersection as exceeding saturation. This corresponds to initial curative treatments, where the hope is to relieve the congestion altogether. The objective based on these definitions is either maximizing throughput or minimizing damaging queues.

Motivating Conditions

Practitioners are usually stretched thin enough such that they require conditions to deteriorate to an action point before they are compelled to do something. The respondents varied in the conditions that would compel action. Several respondents took action when observing operation during routine reviews of new signal timings. All the respondents agreed that observing an unexpected residual queue would compel action, especially if it grew to the point of causing gridlock or a safety problem, and one mentioned accident experience as a motivator. Most reported that citizen complaints were often the trigger that compelled action. One responded to requests for assistance from state DOT district offices, where citizen complaints were received.

Strategies Taken, and Their Objectives

The first step for all respondents, after confirming that there was a problem, was to first adjust the green splits to minimize residual queuing. When this didn't work, the respondent would determine that the intersection was oversaturated and start to consider more specialized strategies. There was no particular sequence to the strategies. Some respondents did not distinguish between the objectives of maximizing throughput and managing queues, while others specifically changed strategies when it was no longer possible to provide sufficient throughput to avoid damaging queues. Thus, some strategies are oriented towards maximizing throughput in the hopes of minimizing damaging queues, while other strategies were reserved for situations where that objective could not be reached. We should note, however, that even though these strategies are divided into the two groups, there is some crossover between them. Most respondents did not make the distinction between throughput and queue management.

We should also note that none of the respondents used optimization tools currently available as part of designing or implementing these strategies, and only two used any form of simulation as a means of evaluating the strategies. All respondents felt that each situation was unique enough to defy the use of a specific sequence of analysis steps. All respondents also felt that improving the problems required the sum of multiple small improvements. Finally, all respondents agreed that the problems and solutions were best identified during direct field observation.

Throughput maximization strategies included:

- Observe the intersection on the ground. Congested conditions cannot be reliably predicted by current models, even simulation models. No current generally available tool is adequate for optimizing timing in congested conditions, particularly at the intersection level.
- Fix problems starting from the most downstream bottleneck.
- Run intersections close together out of one controller, if at all possible. This will provide a tighter level of coordination than coordination timing. Diamond interchanges are the most common example of this strategy. This strategy is mentioned both in the context of throughput maximization and queue management.
- Make sure that lane striping makes most efficient use of pavement. Adding lanes to congested movements has more potential for improvement than any modification to signal operation.
- Lag heavy left turns. This prevents the queue on the through lanes from starving the left turn, which can lead to the left turn spilling out into the through lanes.
- Consider shortening the cycle. Several of the respondents have noted that long green times result in lower departure density. For example, in Oakland County, the SCATS system would measure declining saturation on approaches with

essentially unlimited queues when the greens were too long. One respondent described the "slinky effect", where cars too far from the signal leave a longer headway, and then speed up to try to fill it in, leaving a larger gap behind them. This respondent felt that the densest 15-second period after start of green was the second period, and most respondents agreed with this observation. He noted that greens longer than that required to empty the mainline back to the start of turn bays would serve fewer cars, because cars in the mainline would leave gaps in the main-lane traffic after entering the turn bays. The author has observed qualitatively that headways increase with green times longer than about 30 seconds. These observations are closely tied to maintaining maximum throughput by serving the portion of the waiting queue that can be served at the highest density.
- Consider lengthening the cycle. When cycles are too short, phase-change lost time eats up the efficiency, and required fixed minimum intervals consume a greater percentage of the cycle. This suggestion conflicts with the previous one, and one respondent suggested that the optimal cycle is the "least lousy" balance between these competing objectives. Another respondent termed this "finding the *right* cycle". All agreed that cookbook cycle calculation methods were of little value in congested conditions.
- Consider serving phases more than once in a cycle. One respondent noted that the more heavily movements are imbalanced, the more likely it is to benefit from serving the major movement more often in the cycle, for shorter periods. This allows the minor movements to fill up sufficiently to make use of the minimum time those movements require, without going to a long cycle on the major movement that reduces throughput.
- Consider the effect of buses. Buses tend to consume a whole cycle, short or long. Thus, a shorter cycle on bus approaches usually improves throughput by increasing bus capacity and thus minimizing their effect on throughput. This is not necessarily related to bus priority, or to the question of upstream vs. downstream bus stops. Bus priority usually works within the coordination context, and often the question of relocating bus stops is out of the hands of the practitioner. But by increasing the number of cycles per hour, the respondents felt that capacity could be improved without having to address those institutional issues.
- Minimize the effect of pedestrian movements.
- Seek all possible available green time. Several of the respondents will use complicated controller features in order to allow normally sequential movements to operate at the same time to the extent physically possible. Any green time that can be found from an uncongested movement should be routed by whatever means to a congested movement. This green time may be found by use of inside clearance periods (such as at diamond interchanges), or by carefully balancing the approach splits or omitting unneeded phases.
- Consider the congested and uncongested movements separately. Two respondents mentioned that the strategy of using a shorter cycle for congested movements to minimize the slinky effect and to maximize stop-line density could be coupled with using long cycles on the minor movements to maximize vehicular throughput on phases controlled by pedestrian minimums. This leads to running congested phases twice in each cycle (or running minor phases every other cycle).

- The lighter the pedestrian demand, the more important pedestrian signals and pushbuttons become. The presence of the pushbutton allows the signal timing to routinely ignore the pedestrian minimum, except in those circumstances when a pedestrian places a call.
- Prevent actuated short greens. On approaches with high truck volumes, a truck may leave a larger-than-normal gap, which can cause the signal to gap early and terminate the green. On a congested approach, this reduces the throughput of the approach by not giving them their normal share of the cycle. Set up detection to prevent accidental early gap-outs.

Queue management strategies included:

- Cut down left turn splits, in the hopes of encouraging left turners to make their maneuver elsewhere.
- Ensure that coordination between close intersections remains effective, even during transitions, preemption recovery, etc. Even one failure of coordination can cause a queue to spill back through an upstream intersection, which cascades into a larger problem that may not recover during the peak period.
- Balance queue length for conflicting approaches.
- Adjust splits to prevent a queue from blocking unrelated upstream movements. These include traffic crossing the arterial at an intersection upstream from the congested intersection, and backing a queue up a freeway exit ramp onto the mainline.
- Switch from a fill-then-empty approach to a continuous flow approach (see below). This usually requires shifting to a long cycle, and usually causes serious queuing in places where the queuing does less damage.
- Do whatever it takes within the control equipment to prevent queues from multiplying by extending through upstream intersections.
- Coordinate upstream signals to control flow into the congested intersection.
- Coordinate downstream signals to ensure that they can handle the flow coming out of the critical intersection, without ever queuing back into the critical intersection.

One respondent presented a strategic concept that bears special mention. He made the distinction between two general strategies:

- Fill then Empty. In this approach, the signal timing is adjusted to empty only the portion of the storage that can feed the stop line with the highest density of traffic. This is a throughput-maximization strategy. This approach usually requires shorter cycles, to make sure that green times are controlled to ensure the greatest possible density at the stop line. One respondent (not the one who suggested this dichotomy) noted that as cycle length goes up, the number of fill and drain cycles goes down, increasing the load on each.
- Continuous Flow. Here, the signal timing is adjusted to keep the approach traffic moving, even at the expense of queue length on minor movements, and even if throughput suffers. More specifically, green time is added to the congested movement under consideration, which created a longer cycle without adding green time to the other movements. Thus, the capacity of other movements declined. This was done to prevent traffic from backing onto freeway mainline, and would have to be considered a queue management strategy. This approach

implies longer cycles than would normally be effective when maximizing throughput. During the interview, the respondent and the interviewer noted that this strategy is probably narrow in its application, and many conditions may mitigate against its use. For example, adding green time to the congested movement in large doses may well shift the bottleneck to the next downstream location. The example of this strategy is the third alternative implemented at North Carolina 54 at I-40, described in the previous sections.

Demons

One respondent also presented a series of "demons". The demons were those influences that forced signal timing into states less able to address congestion. These included:

- Longer and longer pedestrian clearance intervals. Not all signal controllers effectively address long pedestrian clearances within coordination, and in many cases long pedestrian clearances drive the cycle length upward, often to excessive levels.
- Longer vehicle clearance intervals. It was acknowledged that motorists will take the clearance they need, and clearance intervals that are too short may cause conflicts. But the respondent was noting longer and longer red clearances in particular. He felt that a red clearance long enough to clear a vehicle all the way across the intersection was excessive, given the lost time cushion for traffic starting up on the crossing movements. He cited one example where the additional red clearances added up to 7 seconds, which is 5% of a 140-second cycle. *Because the minor movements were already at working minimums, all of this time had to come from the main-street movement.* This last point is the critical point: Many who promote the longer clearance intervals assume that those increases clearances come from their associate movements. This is not the case. Because the green times for many of those movements are already as short as possible in congested intersections, the increases around the intersection usually come from only one or two major movements, where they have a significant effect. (The authors note recent research efforts that suggest little long-term improvement from increasing red clearances without due consideration. These are noted in the previous chapter.)
- Too much dependence on detection. At locations where long cycles are used in the hopes that minor movements will gap early, a malfunctioning detector can cause significant unused green time. Shorter cycles minimize the effect of detector malfunctions.
- Too many phases. The more a cycle is divided into smaller components, the less green time each component will get.

Tactics and Controller Features

The respondents used a range of system tools and controller features to support their objectives at saturated intersections. One respondent specifically avoided the use of esoteric features as a matter of policy, because of the concern for maintaining the implementation of those features with maintenance forces who don't understand them. That leads to a general recommendation from all the respondents: *All specialized controller features used in congested conditions should be defined as straightforward features, rather than a combination of otherwise unrelated special features that is hard to comprehend and maintain.* One example is a straightforward feature for serving a phase more than once in a cycle. There are many approaches to doing this (including the method shown in Example 4), but if some agencies are reluctant to overwhelm their technicians with the use of such features, then the solution is to make the features available without overwhelming the technicians.

Overriding SCATS Control. One respondent had a SCATS system running the signals in his jurisdiction, and that provided a means of metering traffic into congested intersections. When the queue from a congested intersection backed through an upstream intersection, the normal SCATS response was to increase green time on the congested approach. Given that the congestion was caused by the downstream intersection, the result was wasted green time, with crossing movements left unserved. He used a special detector to sense the queue approaching the upstream intersection, which triggered SCATS to reduce rather than increase traffic on the through movement feeding the congestion. This would allow minor movements to be served.

The respondent's experience with SCATS brings up a more general issue of adaptive control in saturated conditions. The objective of current adaptive control systems is to adjust timing so that the critical intersection hovers at a high level of saturation, typically about 90%. Adaptive systems respond to the intense detection response associated with a standing queue by adding more time to that movement. They do not balance queue lengths, nor do they have any mechanism for addressing damaging queues. When demand increases, they all seek longer cycles, even though this may not be the preferred strategy to maximize throughput. As with off-line optimization software, none of the adaptive optimization algorithms in common use seek to maximize throughput or minimize damaging queues as an objective function. For this reason, the respondents generally did not consider adaptive control as likely to be effective in congested conditions, particularly when considering single or small groupings of intersections. Only one respondent manages an adaptive system routinely, and as the above paragraph explains, he sometimes has to work around that system's algorithm.

Proponents of adaptive control suggest that these systems may delay the onset of congestion even if they are less effective once the congestion arrives. As a matter of opinion on the part of the researchers, this varies depending on the adaptive system being considered. Systems that minimize queue formation (as a byproduct of maximizing throughput) will prevent congestion most effectively, and those considering adaptive control would be encouraged to understand the objectives of the optimization process used by the adaptive system.

Phase Reserve. Several respondents manipulated their controllers to provide green signals more than once in the cycle, or to alternate movements to every other cycle. For example, the author once used a cycle twice the coordination cycle to serve split-phase side-street movements on alternating cycles. These movements were light, and serving them on alternate cycles made better use of the minimum green time, which was controlled by pedestrian clearance time.

One respondent suggested that the more heavily movements are imbalanced, the more likely the intersection will benefit from serving the major movement more often in the cycle, for shorter periods.

Minimize Pedestrian Effects. When pedestrian demand is heavy, then all respondents agreed that it had to be included in the signal timing explicitly. But when pedestrian demand is light, the temptation by many agencies is to avoid installing pedestrian signals and pushbuttons. As mentioned above, the presence of signals and pushbuttons allows pedestrian movements to be routinely ignored unless there is a pedestrian call. The traditional approach to this is to provide coordination timing that violates the pedestrian intervals. When the pedestrian intervals are called by a pushbutton actuation, they override the coordination timing to serve the pedestrians. The intersection then allows the signal to transition back to coordinated operation. If this happens only rarely, the loss of coordination has negligible overall effect. In this case, "rare actuation" is considered to mean an actuation no more often than once in several hours.

Not all controllers handle pedestrian override as defined above. For example, some controllers transition slowly, while others won't allow coordination timings that violate pedestrian intervals to be installed in the first place. One respondent suggested a workaround. He sets the split to allow for the pedestrian intervals, but then sets the maximum green time (MAX) in use during that pattern at the normal green time needed by vehicles. The phase will normally max when the pedestrian intervals are not called. If all the MAX times are set to their desired vehicle splits, then the MAX times will control the splits of the non-coordinated phases. Some controllers inhibit MAX during coordination, and this must be disabled. When the pedestrian intervals are called, the intersection runs according to the coordination timings, with no loss of coordination and no requirement for transition.

Some controllers allow MAX settings by pattern, and some by time of day.

Detector Switching. Several respondents used detector switching to turn off detectors that cause an approach to extend for density values that are too low, and for turning on detectors that serve greens that overlap with other movements to increase efficiency. An example is to switch off a detector in a right-turn bay if the complementary left turn will be served next (i.e., if it has a call). This requires using special controller logic.

Special Controller Logic. One respondent aggressively used auxiliary logic functions within controllers to achieve specific objectives. For example, when a left turn lags, he uses an auxiliary function to extend the complementary through movement even if it has no call.

Another respondent used controller logic to circumvent normal barrier cells. The logic controls in their software allow them to assign a logic output to a load switch, and then control that logic output by any combination of normal controller conditions.

Simultaneous Gaps. Once one of a lagging phase pair gaps, the controller will not extend that phase until the next cycle, even if it is holding in green waiting for the other lagging phase pair to terminate. This prevents alternating widely spaced cars to hold opposing through movements in green despite low density being served.

Dual Entry. This feature causes compatible minor movements to both turn green even if only one of the phases was called. For example, if the side street has a call on Phase 8, the controller will also serve Phase 4 along with Phase 8. Without invoking this feature, Phase 8 would be served while Phase 4 would remain red, unless a car placed a call some time later in the Phase 8 green period, at which time Phase 4 would go green and time all its minimum intervals. Dual entry forces side-street compatible pairs to be served at the same time.

Dynamic MAX. One respondent mentioned this feature, though it was not favored by all respondents. If a phase maxes out in successive cycles, the MAX time will be increased by an increment. This will continue for each successive maxing out until a maximum MAX is reached. Some controllers revert the max time back to the normal setting on the first phase that gaps out instead of maxing out, while others will subtract increments on successive gap-outs until it backs down to the normal setting.

A variation on Dynamic MAX is adaptive split operation. One respondent reported the desire to have coordination timings adapt to the presence of detection throughout the green period. In other words, when cars are detected right through to the end of green repeatedly, the controller would adjust the coordination timings accordingly.

Volume-Density. Use variable gap to shorten gap settings quickly to more aggressively find a gap in the approach traffic. Also, use variable gap on stop-line detection zones to prevent a truck-induced (and therefore larger) gap from causing a short green.

Desired Controller Features

Most of the desired controller features are available from one or two manufacturers, but they are not part of the NEMA standard and therefore not universally provided. Some require logical functions external to the basic NEMA ring rules. Desired controller features include?

- Eliminate barrier cells. Currently, the NEMA ring diagram provides a barrier across the two rings of sequential phases. It is more flexible to set a phase sequence in a ring, and then declare which phases in one ring are compatible with which phases in the other rings. One responded uses this capability with lead-lag operation that is forced by geometry, to prevent opposing left turns from operating at the same time without imposing a particular sequence.
- Allow uncomplicated and straightforward phase reserve. Several respondents support this request.
- Variable gap time, by time of day.
- Floating force-off, by phase. This accomplishes a similar objective as using MAX times to control splits (see previous section, under Minimize Pedestrian Effects).
- Right-turn detector logic, to allow adjacent through movements to terminate when the next phase is the complementary left turn. This would inhibit the controller from holding the through green on the basis of right turn detections only, when those rights turns could be served during the complementary left turn coming up.
- Split-phase operation based on queue detection. Split phasing is more efficient when options lanes are provided and when turning traffic exceeds through traffic. This should be evaluated cycle by cycle.
- Variable walk time based on push button input.
- Pedestrian cross-switching and head starts by time of day. This feature would allow pedestrian movements to overlap with non-conflicting left turns, or with left turns that do not conflict with the pedestrian movement during the first portion of the pedestrian interval.
- Eliminate the requirement to have a coordinated phase in each ring.
- Omit minor movements by time of day.
- Measure detector on time from the first time it went on, regardless of phase color or service, as a measure of residual queuing. This was one of the few

suggestions by a respondent that would measure the boundary between heavy traffic and building residual queues.
- Monitor left turn detector to determine, regardless of phase color, when a queue for a permitted (not protected) left turn is never served. This applies to cases where no protected left turn is provided to serve a left turn bay. The elimination of the protected left turn is often done by time of day, and in some cases corresponds to a left-turn prohibition. A left-turner that is unwilling to turn permissively my become trapped in the bay in this circumstance.
- Drop protected left turn when left turn volume does not warrant the separate phase. This could be coupled to the previous item to determine when to turn the protected phase back on.
- Provide a monitor of splits, showing when phases gap or max out. This is a feature available in some current systems.
- Adaptive split, which provides the same general feature as adaptive max but is applied to coordination timings. When a phase consistently sees detection through the end of green, adjust splits upward for this phase and downward for phases with less utilization, etc.

III. Evaluation of Strategies

In this section, strategies that were subjected to simulation, field testing, or both are described in detail. For reasons of project budget and scope, not all strategies were subject to such testing, nor is testing required justification for all strategies. For those strategies, such as phase reservice where the benefits can be demonstrated directly, further investigation was not conducted. The sections below therefore focus on four issues and strategies with benefits that defy current beliefs or practice, or that are not fully supported by existing research or intuitively obvious. They include:

- Long green times and cycles in the congested regime.
- Using a single controller to operate two nearby intersections, when the needs of coordination between the intersections deserve higher priority than what might be provided by coordination. The example used in this work is a diamond interchange.
- The relationship between bus capacity, network topology, and cycle lengths in terms of bus queuing.
- Leading and lagging left turn phase sequences.

Long Green Times and Cycles

Introduction

In addition to the evaluation of strategies suggested previously, the review of the state of the practice revealed an effect that has not been previously studied in the context of saturated flow. The interviewees reported that stop line flow seemed to decrease after the first half minute or so of green, even if the flow is still being fed by an emptying queue. This was reported both as a qualitative effect, where one interviewee suggested that the second 15-second interval following onset of green showed the highest apparent flow, and as a quantitative effect, where an interviewee noted that his agency's SCATS system had to be detuned on approaches with unlimited queue because it tended to note a drop in density and incorrectly interpret it as a drop in demand.

A review of the literature reveals only two studies that look into the stability of stop-line flow directly. One was performed by Stan Teply for the Canadian Capacity Guide. Teply's data showed that flow peaked at around 2000 vehicle/hour in the interval between 15 and 30 seconds from start of green, and then after 30 seconds diminished to a value closer to 1800 vehicles/hour. These results would seem to confirm the impression of the interviewees.

(There was additional work done as part of the supporting work surrounding NCHRP 3-66, where Bullock and others looked at flow with respect to the end of green. But they were primarily studying the flow of cars at the point where the signal controller gapped out, not at the flow of vehicles being fed by a standing queue. This work was reported in the Transportation Research Record 1978, published in 2007.)

Karan Khosla, in a Master's Thesis at the University of Texas at Arlington (and partially summarized a paper also included in TRR-1978), studied the matter at four intersections in the Dallas/Fort Worth metropolitan area. Khosla found that the flow increased with time into green through about the first 20 seconds (the flows were reported in 10-second intervals), and then stabilized until the final one or two 10-second intervals, where it increased (presumably as an effect of rushing the clearance interval, and considering that not all green periods were the same). At two of the four locations, the flow diminished starting with the interval at 40 seconds into the green (excepting the final interval). Interestingly, the two intersections that showed a reduction had auxiliary turn lanes for both left and right turners, though the data were not collected from the lane adjacent to the turn lanes. In fact, the major weakness of the work is that data were collected for a single center lane only, and not for the whole approach.

Two causal mechanisms have been hypothesized to explain the effect of diminishing stop-line flow from an emptying queue, assuming it exists. One is that turning vehicles may be trapped in a long queue in the through lanes. They will depart the through lanes for the turning lanes during the green for the through vehicles, thus lessening the flow on the through lanes. A more severe cause of this effect is an approach that widens to additional through lanes as it approaches the intersection. As the departing queue works its way back to the narrower section, the stop line can only be fed by the cars stored in that narrower section, which would starve the capacity of the wider section at the stop line by some amount.

The second mechanism might be that drivers are no longer able to respond to the displayed green light, because their position in the queue puts them too far to see the signal. For example, if the flow begins to drop after 40 seconds from the start of green, the vehicles being serviced at that time were about 20 cars back from the stop line, or about 400' from the intersection. If the drivers cannot see the signal, their perception-reaction time may no longer overlap with the vehicles in front of them as characterized originally by Greenshields. If they are reacting instead to the brake lights of the cars in front of them, or to movement in an adjacent lane, some of the perception-reaction time might be added back into their initial headway. As they speed up to try and close this larger gap, they create shock waves as with a freeway queue resuming its normal speed in a bottleneck. It was unknown at the outset of this study whether this effect would reduce the stop-line flow measurably.

The relevance of this issue is significant for saturated networks. One operating assumption common among practitioners is that intersection total throughput increases as cycle length increases, because the phase change lost time, which is constant, diminishes as a percentage of the cycle. This assumption apparently conflicts with the anecdotal observations summarized above, and with Teply's research and partially with Khosla's research.

The common practice of increasing cycle length to increase capacity is based in part on the assumption that saturation flow remains constant once the initial lost time has been accommodated, and the effect of traffic departing from through lanes for turn lanes is ignored.

If this assumption turns out to be false, or if the effect of departing turn traffic is significant, then the common practice might not have the effect of increasing throughput. As we have established previously, maximizing throughput is the primary objective function for optimizing saturated intersections.

A comprehensive study designed to model the change of saturation flow through long greens is beyond the scope of the project. However, the researchers evaluated one location with field studies and simulation to evaluate the effect of the two postulated causes, and considered the impact of these effects on throughput as the basis for developing a throughput-based cycle length recommendation to practitioners.

Study Site

One intersection in Northern Virginia provided an excellent opportunity to evaluate this issue. Route 28 in Fairfax and Loudoun Counties connects I-66 in the vicinity of Manassas to Dulles International Airport and Route 7, which is the primary arterial facility running parallel to the Potomac River in the western suburbs of Washington, DC. Route 28 serves many large traffic generators, including the National Reconnaissance Office, the headquarters campus of AOL Time-Warner, Verizon's regional headquarters (in the former MCI-Worldcom campus), and a host of others. The Virginia Department of Transportation has been systematically replacing at-grade signalized intersections along Route 28 with grade-separated interchanges, elevating Route 28 to expressway status.

The intersection of Route 28 and Frying Pan Road is one of the few remaining at-grade intersections controlled by a traffic signal. The next signalized intersections both upstream and downstream from Frying Pan are about five miles away, and in both directions major facilities freely interchange with Route 28 before reaching those at-grade intersections. Thus, demand from upstream is little affected by traffic signals and there are no constraints on traffic downstream from the location.

Demand at the intersection is very high, about 875 vehicles actually served on the northbound approach in the peak 15 minutes of the study. This corresponds to an hourly flow of roughly 3500 vehicles per hour. Note that these vehicles crossed the stop line, and the queue at the beginning of the field study exceeded four miles. Thus, actual demand is significantly higher than 3500 vehicles per hour during the morning rush when the study was conducted. Route 28 carries this traffic on three basic lanes in the northbound direction, adding a right-turn lane about 500 feet upstream from the stop line. Frying Pan Road is a TEE intersection with no northbound left turn.

Signal operation at the intersection was fixed, by observation. Green times were measured to be 180 seconds in a cycle of 270 seconds.

Figure 3-1 shows the intersection of Route 28 and Frying Pan Road.

Figure 3-1. Virginia Route 28 at Frying Pan Road (North is up, which is the direction under consideration).

Field Data

The researchers developed a software program that allowed a field observer to record the time each vehicle crossed the stop line. The observer first noted (by key press) the start of green, and then pressed the 1, 2, or 3 key each time a vehicle in the corresponding lane crossed the stop line. Key-press times were recorded to the nearest millisecond. The observer was located on the downstream side of the intersection in the east right of way, on a small embankment that provided a good view of the northbound approach. The software recorded the time of each key press in a comma-delimited text file which was imported into Excel for subsequent processing.

The observation method is subject to systematic errors based on the skill of the observer. The observer noted potential errors of up to around a half second in recording the crossing time of vehicles. The observer reported no expectation of actually missing any vehicles greater than one or two per cycle (well under 1%, given the average flow in each cycle of approximately 270 vehicles). Thus, the data cannot be used to develop or evaluate car-following relationships between individual pairs of vehicles, but this was not the objective of the field study.

In order to filter out the noise induced by any flaws in the observation technique, vehicles were grouped in ranks of five vehicles and the headways between them were averaged. An error of half a second would be one quarter of a typical two-second headway, but when averaged across a succession of five vehicles, would be closer to 5% of the true headway, on average. Given that no more rigorous data collection method was attempted or allowed by project resources, a more rigorous evaluation of data-collection error was not attempted. However, as we will see, each rank of five vehicles in each lane represents 40-55 actual headway measurements, and data collection error is assumed to average out to become unimportant to the conclusions of the study.

Data were collected over 11 consecutive cycles during one morning peak period in March of 2008. Flow in each lane ranged from 58 to 101 vehicles during the green. Initially, the data were only evaluated for the first 60 vehicles crossing the stop line in each lane. This decision was made for two reasons. One is to avoid having to deal with more than one instance where a lane did not serve that many cars during the study period. The other was a recognition that the first 60 cars were all that were needed to test the two effects being studied. Using conventional assumptions, 60 vehicles would require slightly less than 120 seconds of green time to serve, which is already an extraordinarily long green time.

The headways between the first 60 vehicles in each lane were first evaluated to test for correlation to the cycle number. Such correlation would indicate that conditions changed during the course of the data collection, which would need explanation. The data showed that the headways of vehicles from queue positions 40 to 60 were longer during the last three cycles of the study than during the first eight. These cycles were at the end of the AM peak period, and the explanation seems to be lessening demand and gradual clearing of the residual queue. Therefore, headways after the first 40 vehicles were ignored for the last three cycles. As a result, 11 cycles were used for the first 40 vehicles crossing the stop line, and 8 were used for the next 20. A total of 1800 vehicle headways were included in subsequent analysis. Figure 3-2 shows these headways plotted by lane and cycle number. Lines connecting dots represent successive vehicles.

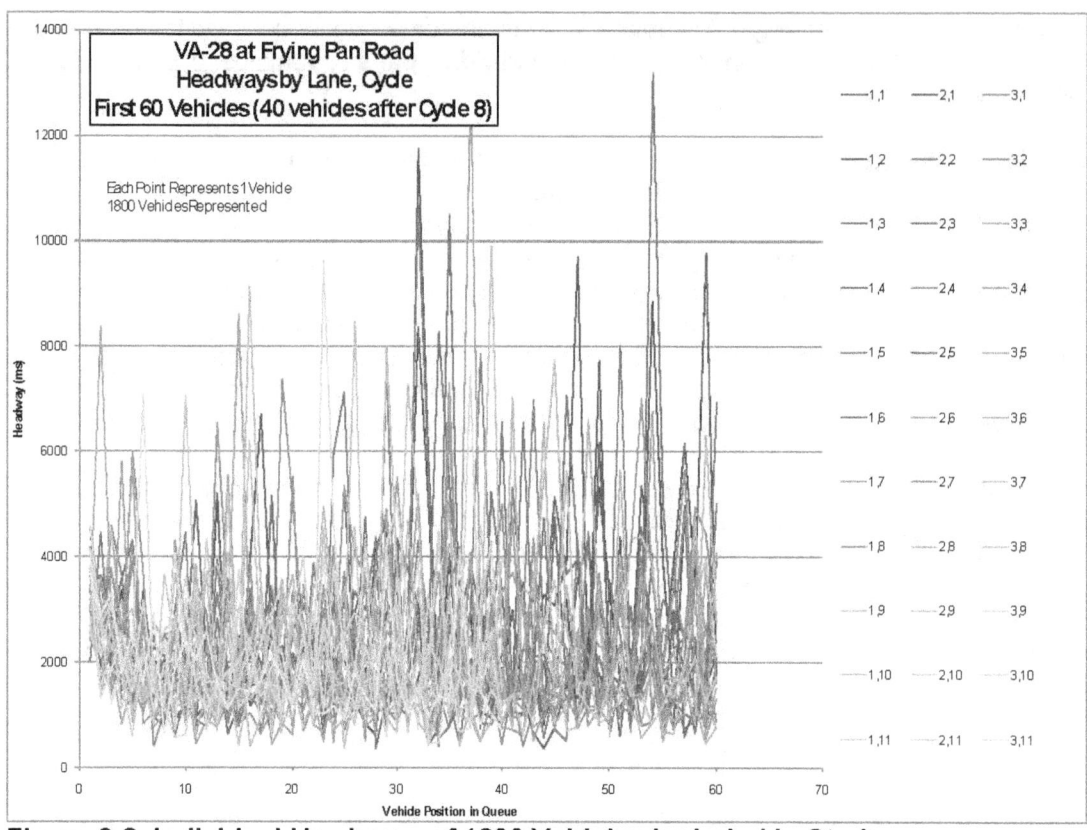

Figure 3-2. Individual Headways of 1800 Vehicles Included in Study.

The 1800 vehicles included do not show correlation by cycle numbers, as is shown in Figure 3-3. Note that for all evaluations of headway, the first five ranks of vehicles were ignored to eliminate the effects of startup lost time. This effect is clearly visible in Figure 3-2.

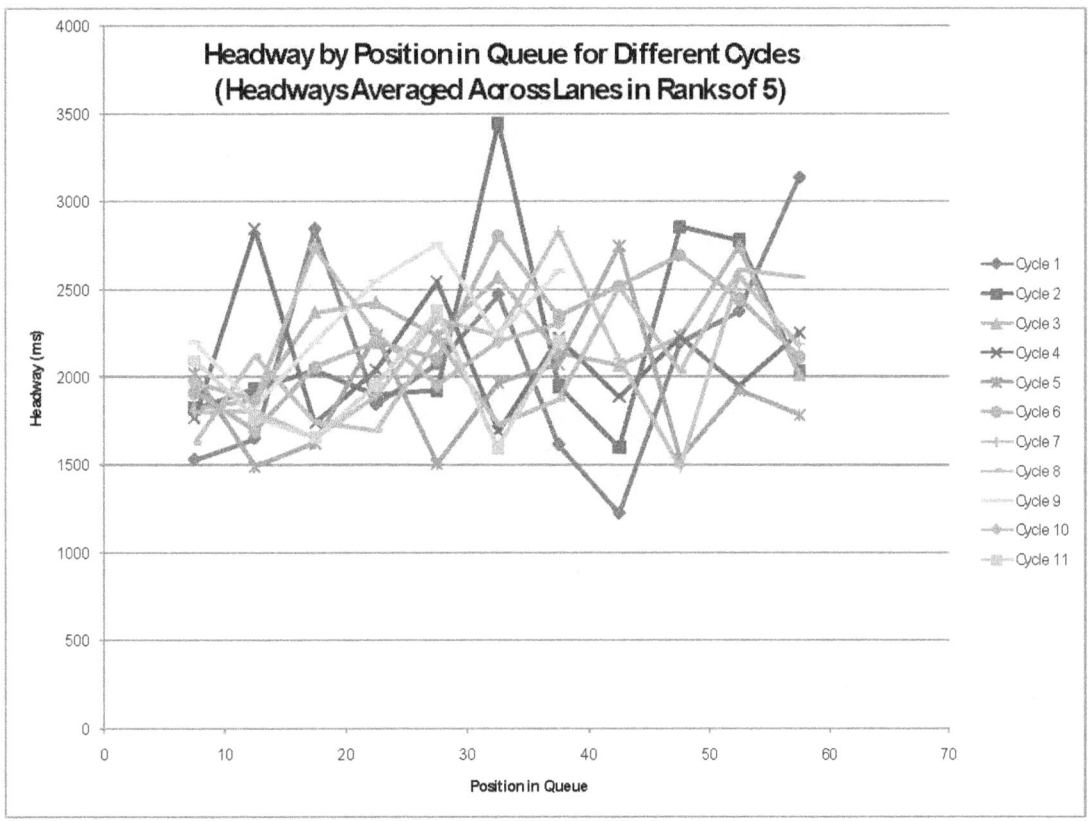

Figure 3-3. Headways Compared to Cycle Number.

Another way to display these data is to show vehicle stop line crossing by time into green. Figure 3-4 shows time-into-green crossing by lane and cycle. The thick line on the chart shows the Greenshields model with an assumption of four seconds lost time and a saturation flow of 1900 vehicles per hour of green. One interesting immediate observation is that the Greenshields relationship using currently accepted numbers proved to be an ideal not often exceeded by the field data.

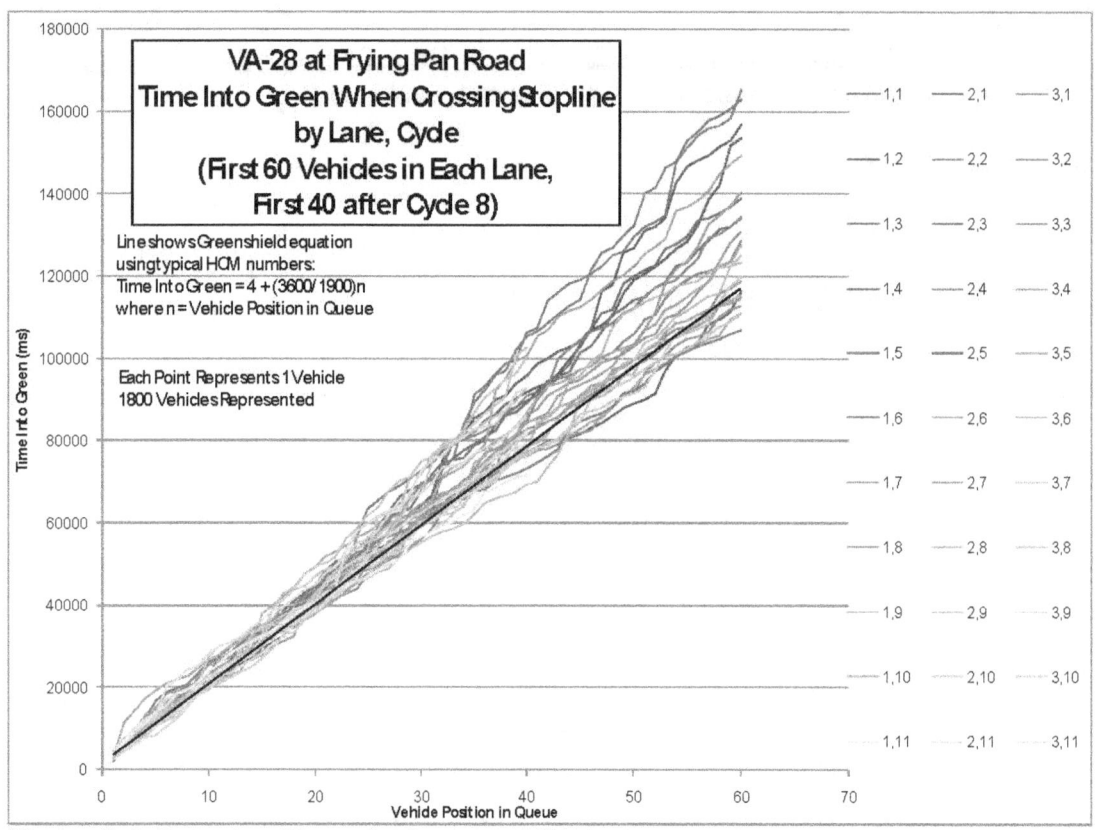

Figure 3-4. Crossing Times by Lane and Cycle.

The results of primary interest are shown in Figure 3-5. This chart shows average headways by lane (and across all lanes) averaged in groups that include five ranks of vehicles, from the 6th to the 60th vehicle in the departing queue (40th vehicle for the last three cycles).

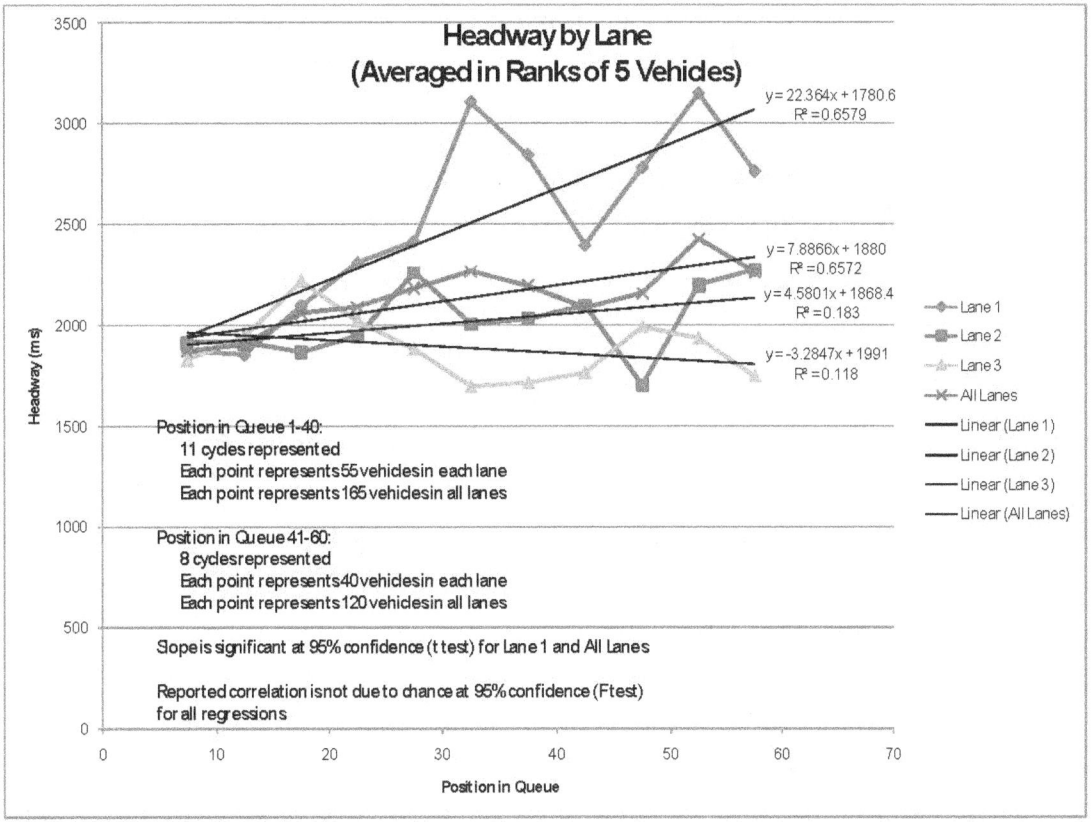

Figure 3-5. Headway by Lane.

Analysis of Field Data

Linear regressions were conducted for headways (in ranks of 5) for each lane and for all lanes. The regressions provide an assessment of changing headway (the slope of the line) and correlation between queue position and headway. Correlation alone is insufficient to show that headways increase, statistical testing is also required on the coefficient of the position in queue parameter (which is the slope of the line) to determine that the slope is significant.

Thus, a two-tailed test of the t statistic was performed on the slope parameter. The linear regression for the headways in the right through lane was:

$$H = 22.4(Q_{pos}) + 1781$$

where

H = headway (ms)

Q_{pos} = position in departing queue (1)

With 11 data points in the regression, the calculated t statistic was 4.16. With 9 degrees of freedom, the value greater than which one rejects the hypothesis that the slope value is due to chance at a confidence of 95% is 2.26. Thus, for the right lane, the slope is significantly non-zero.

Similar calculations were made with the middle and left lanes, but the slopes of those lanes were not sufficiently different from zero to reject the hypothesis that they are caused by chance.

The conclusion is that the right lane saw increasing headways, but the center and left lanes did not. Total headways, averaged across all lanes, did show significant slope, based on the influence of the right lane in the average.

The left lane was an ideal test for evaluating the hypothesis that driver reaction time starts later and adds to the headway beyond a certain distance from the stop line. Yet headways did not increase with position in queue in the left lane. Thus, the data do not support this hypothesis.

The right lane was an ideal test for evaluating the effect of turning traffic leaving the through lane. The right-turn lane is about 500 feet long. Approximately 25-30 vehicles would be expected to fill a 500-foot queue, so one can assume that few right-turners will be in the first 25 or 30 cars because they would have already been able to move into the right turn lane. Thus, we expect to see headways mimic the center and through lanes through a position in queue of 25 or 30. The data do show a slight increase in headway through that much depth of queue, but then jump significantly thereafter to a peak at a queue position of 30 to 35. After that peak, the headway drops back down around the 45th vehicle, and then peaks again at the 55th vehicle.

A possible explanation for this behavior is that the departing traffic empties the queue to a point beyond the opening of the right-turn lane, allowing right-turners to leave the right through lane to make their maneuver. As the right-turners leave the through traffic stream, they leave gaps. Drivers behind them may accelerate to fill those gaps, but are unable to catch up to the gap by the time they reach the stop line. The opening creates a shock wave, and flow increases again, resulting in shorter headways, as the shock wave emerges. After the shock wave passes, headways increase again. This type of behavior is well-known from freeway traffic.

To test this possible explanation, the intersection was simulated to see if the shock wave would be visible, which is reported in the next section.

The middle lane showed growing headways to a greater extent than the left lane, though this was not found to be statistically significant. If it does, however, then this could be explained by some vehicles in the middle lane moving into gaps created by departing right turners in the right lane before reaching the stop line.

Correlation was also tested. Some correlation was seen in all cases, but the correlation coefficients range from moderate to low. The F statistic was used to evaluate whether these correlations could be attributed to chance, and in all cases, the F statistic exceeded the 95% critical value on the F distribution, meaning that the reported correlation is probably not due to chance.

Given the low correlation coefficients, however, it is not recommended that the linear regressions be used as a model to characterize headways.

One question to be evaluated is the effect of the headways on various assumptions pertinent to calculating greens times at saturated intersections. One such is the assumption that the capacity of the green time can be characterized by the equation

$$Capacity = \left(\frac{G}{C}\right)S$$

    where

        G = green time

        C = cycle length

        S = saturation flow   (2)

This equation makes the assumption that saturation flow is a constant. The Greenshields relationship was traditionally

$$G = L + Hn$$

    where

    G = green time required to serve n cars

    L = start - up lost time

    H = average headway

    n = number of cars to be served   (3)

In this equation, L is the amount of lost time used up by the first several vehicles, and H is the amount needed by each car thereafter. Greenshields suggested numbers of 3.8 seconds for lost time and 2.1 seconds for headways. Thus, 2.1 seconds headway, when inverted to flow, constitutes the saturation flow, which is traditionally measured as the flow required from the 4th through the 10th vehicle in a departing queue. The inverse of 2.1 seconds is 1714 vehicles per hour. Greenshields has been updated with new measurements of saturation flow, and the Highway Capacity Manual now recommends a value of 1900 vehicles per hour of green. This corresponds to a headway of just under 1.9 seconds per vehicle.

Also, most practitioners estimate lost time at four seconds, which is a number considered to be reasonable in the absence of more specific data.

Thus, the 60th car in a departing platoon should reach the stop line in 118 seconds, which is 4 seconds lost time plus 1.89 seconds headway multiplied by 60 vehicles.

Does the field data support this calculation? Figure 3-6 shows the average time required to serve cars at each rank in the departing queue. The numbers were calculated by averaging the time into green for all the cars in that rank and lane (or across all lanes).

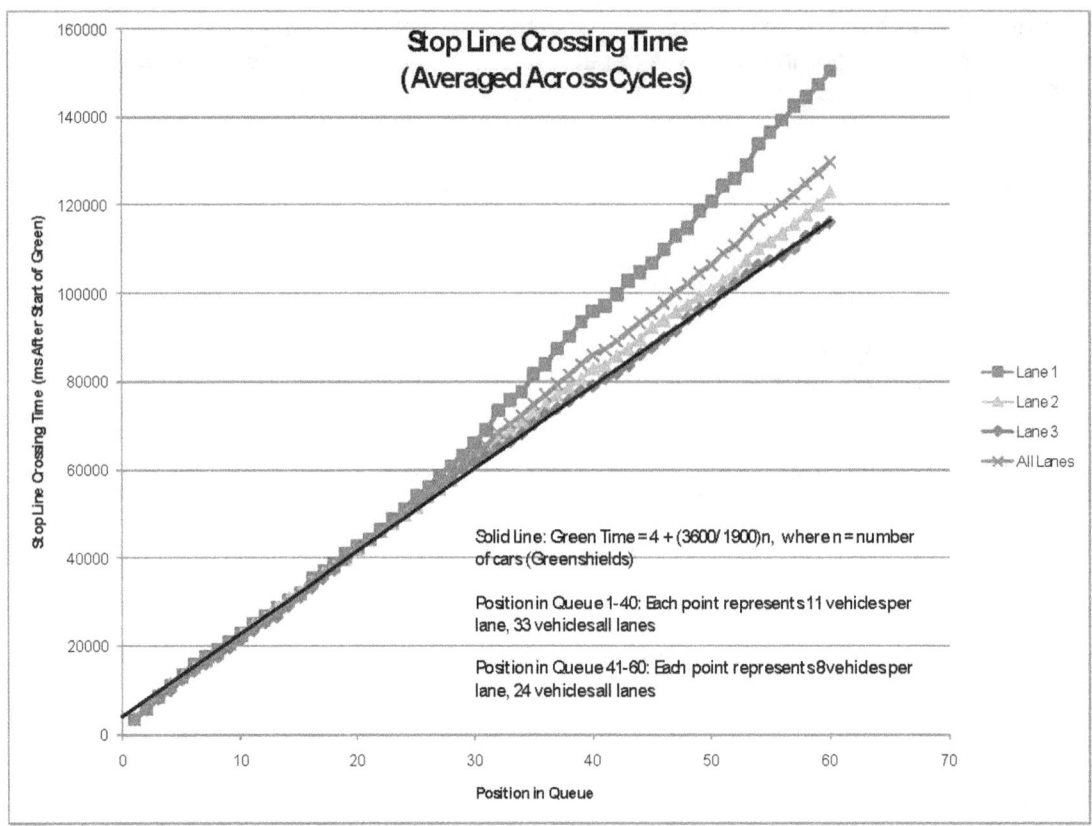

Figure 3-6. Time Into Green by Rank and Lane

The solid line in Figure 3-6 illustrates the Greenshields model with modern parameters.

The left lane corresponded very closely to the Greenshields model, and the middle lane was close, with an accumulated error of about five seconds by the 60th rank of vehicles.

The right lane, however, fell behind the ideal by around 35 seconds. Note, however, that the right lane closely corresponded to the ideal through about the 25th rank. This is the point where headways spiked in response to the queue clearing back to the point where right turners would again be in the through lane.

Field Data Conclusions and Further Discussion

The field data constitutes a single condition and location, and consequently cannot be used to construct models. But the data do show that lanes adjacent to turning lanes will show significantly increased headways and reduced stop-line flow once the queue has cleared to the upstream end of the turning lane. The field data showed only the effects of right turners, which always receive green at the same time as through vehicles. We would expect the effect to be more pronounced with left turns, for these reasons:

- Left-turn bays are often shorter than the right-turn bay at the study site.
- Left-turners are usually separately signalized at congested intersections, and thus may back into the left through lane, causing a blockage.

Even the effect of a right lane, however, was sufficient to cause a significant increase in average headway across all lanes, and an accumulated error in predicted required green time of about 15 seconds.

Headways were not affected by turning traffic downstream of the turn lane entrance. *This suggests that maximum throughput is served when green times make use of the ability to feed the stop line with maximum flow.*

We can turn the analysis around. We have been asking how much time is needed to serve a given number of vehicles. To evaluate throughput, however, it is useful to ask how many vehicles can be served by a given green time. For example, Figure 3-6 tells how many vehicles can be served in 118 seconds of green. The average across all lanes was 54, or 162 vehicles in total.

We can also ask, how many vehicles can be served in, say, 48 seconds? From the data, the average across lanes is about 23, or 66 vehicles in total, on average per green period.

If, hypothetically, the 118-second green were two-thirds of the cycle (as is the 180-second green in a 270-second cycle as observed in the field), the cycle would be 177 seconds. The hourly throughput would be 3295 vehicles (20.339 cycles/hour times 162 vehicles/cycle).

On the other hand, if a 48-second green time was also two-thirds of a cycle, the cycle length would be 72 seconds, of which 50 could be operated in an hour. The total throughput for that hypothetical scenario would be 3300 vehicles. The significance of 48 seconds is that this is about the green time required to serve the vehicles that can queue up downstream of the turn lane entrance.

(Note that a 72-second cycle is impractically short at this location because of the very short times that would be available for the other movements. One advantage to lengthening the cycle is that it allows long enough greens so that they can be assigned based on traffic rather than being forced by minimum-green considerations unrelated to traffic volume. This has been temporarily ignored in this discussion to evaluate the cycle-length-vs.-throughput issue directly.)

Thus, by keeping the green time down to the point where only the queue to the upstream end of a 500-foot turn lane was served in each cycle, flow at the stop line is maintained close to ideal saturation and the overall throughput does not decrease. We conclude, therefore, that the common belief that longer cycle lengths can be assumed to result in greater capacity cannot be supported by behavior at (at least one) real intersection.

The actual signal timing at the intersection was 180 seconds of green in a cycle of 270 seconds. The measured flow during the first eight cycles, which were observed to have maintained queue-departure flow for the entire green time, was 1986 vehicles in 2160 seconds, for an hourly throughput of 3310 vehicles. Again, increasing the cycle does not result in greater throughput.

Thus, the reduction in lost time as a percentage of the cycle does not overcome the reduced flow caused by turning vehicles leaving the through lanes in very long greens.

In the next section, we will test this thought experiment with simulation of the three signal timing alternatives.

One final point should be made. This intersection was only affected by a very long right turn lane. Most intersections include both right and left turners, and usually with shorter left turn lanes and many times no right turn lanes. Thus, we would expect these results to be even more pronounced at more typical arterial intersections.

Simulation

For analyzing optional control approaches, VISSIM v. 4.3 was used to provide microscopic simulation. Setting up VISSIM to model the condition with reasonable realism required many iterations and adjustments to the link and connector geometry. In particular lane changes were a problem in queued traffic. Motorists in the real world have prior expectations of congestion and find a way to edge into the desired lane usually before being forced to block traffic for an extended period of time. Even those drivers who do wait to the last minute will usually give up on making the lane change, or will benefit from a charitable driver in the lane into which they are moving, or will force their way in more aggressively than VISSIM was prepared to simulate. VISSIM would simulate right turners, for example, waiting in the middle lane adjacent to the upstream end of the right turn lane for the entire 60 seconds that VISSIM would allow such waiting before eliminating the vehicle from the simulation and reporting an error. To prevent these behaviors, we set up the VISSIM network such that right turners started in the right lane at the outset, to prevent such lane changes altogether. The assumption we were forced to make was that these lane changes would not affect headways at the stop line.

We also provided two parallel links covering the length of the right-turn lane. One link served the left two lanes and the other served the right through lane and the right-turn lane. This configuration made it possible for right-turners to move into the right-turn lane at any point along its length. At the right-turn island gore, the two parallel links connect to a single link serving the three through lanes.

Input volumes were chosen to ensure that the network was oversaturated, with an approach hourly volume of 6000, 750 of which turned right. The 6000 entering vehicles were divided equally among the three lanes. Approach links were made as long as the license of the copy used permitted, about 8000 feet. The queue overflowed the approach links throughout the simulation, but this caused no visible effect at the intersection. The simulation was run for 4000 seconds, and data were collected starting 1080 seconds into the simulation (four cycles). The signal timing was the same as was observed in the field study. The simulation stopped during a red period, so data were collected from ten consecutive departing queues.

Figure 3-7 shows the layout of the VISSIM network, with links and connectors shown only by their centerlines.

Figure 3-7. VISSIM Network, VA28 at Frying Pan Road

We used the Wiedemann 74 car-following model in VISSIM. That model describes the distance, *d*, between cars as

$$d = ax + (bx\_add + bx\_mult \times z)\sqrt{v}$$

where

ax = standstill distance

v = vehicle speed (m/s)

z = a normal random variable in the range of 0..1 with a mean of 0.5 and a standard deviation of 0.15

bx_add, bx_mult = user-defined parameters (4)

The user-defined parameters are suggested in the VISSIM documentation to be the preferred choice for adjusting the saturation flow. Several values were tried, and the resulting headway and time-into-green performed best, based on a visual evaluation of the charts, using a *bx_add* of 2.6 and a *bx_mult* of 3.6. These values compare to the default values of 2.0 and 3.0, respectively. The standstill distance was left at the default two meters.

VISSIM was configured to export raw data for detector zones at the stop line. The data were manually filtered to include only records that indicate a vehicle first entering a detector. These records were then evaluated in an Excel spreadsheet similarly to data collected in the field. Figure 3-8 shows headways as recorded in the simulation. As with the field data, the headways were averaged across ranks of five departing vehicles.

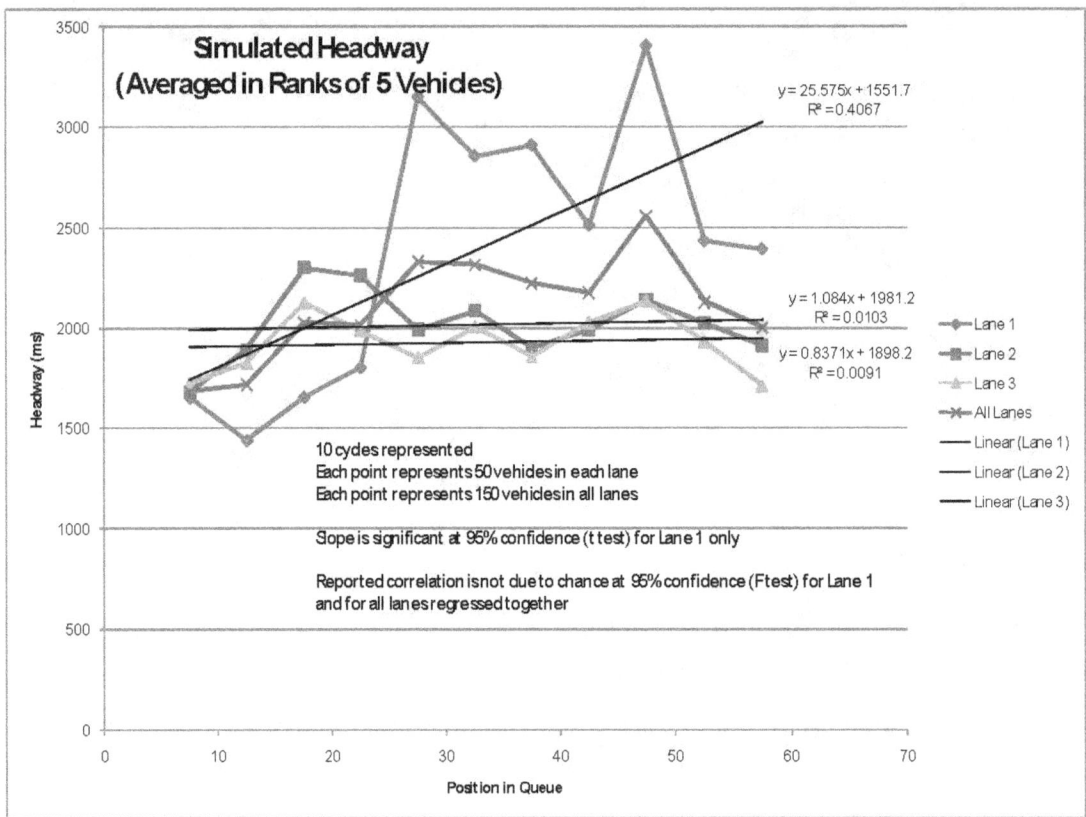

Figure 3-8. Simulated Headways by Position in Queue.

The simulated data show a similar shape to the field data, in that headways remain the same with respect to queue position for the middle and right lanes, but lengthen as a result of departing right-turners in the right through lane. The simulation also shows a sudden increase in headway in the right lane just upstream from the entrance to the right-turn lane, followed by a return to shorter headways, and then followed by another hump. No simulation values were found that matched the field-measured y-intercept of the linear regression for the right lane, but the slope was similar and the range of values were similar. Also, the sudden increase in headways at the upstream end of the right turn lane happened earlier in the departing queue by a few cars. The simulated headways in the right lane were just a bit short overall, but were judged close enough to the field data to provide a reasonable means for comparing alternative strategies.

Figure 3-9 shows the time into green, again with a solid line representing the Greenshields equation with modern values, as pictured in Figure 3-6.

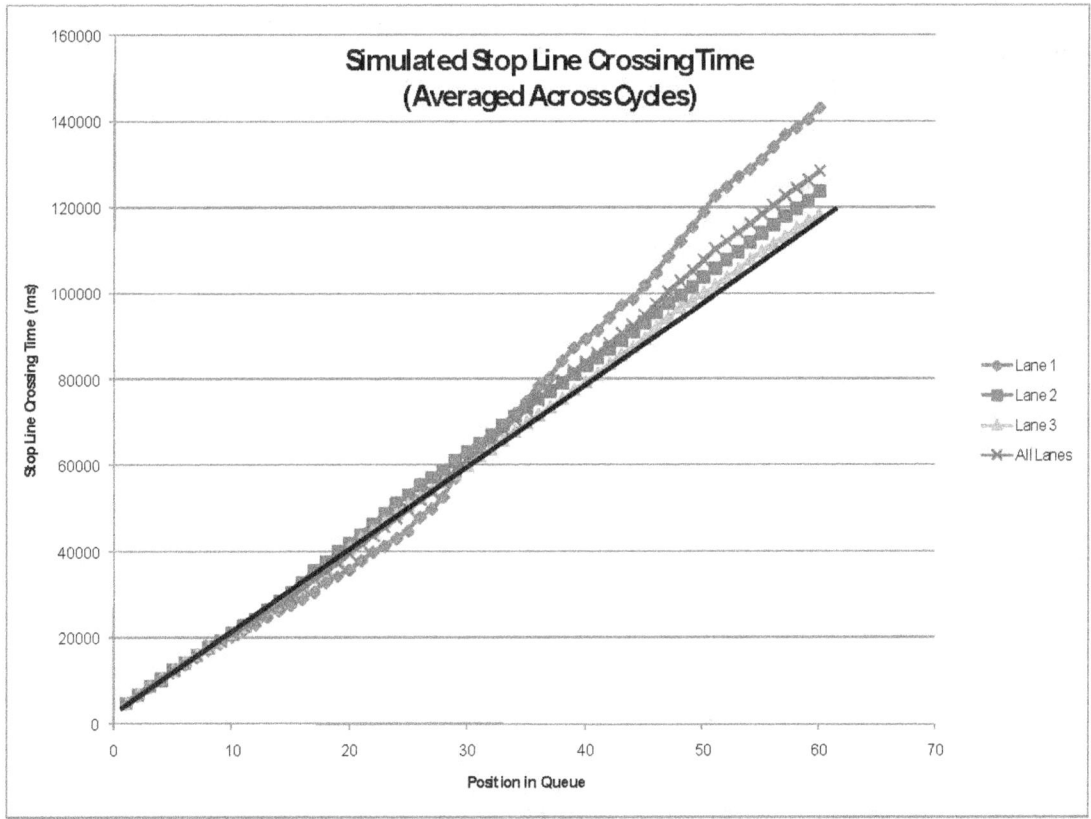

Figure 3-9. Simulated Stop Line Crossing Time

Comparing Alternatives
The main purpose of the field study at Frying Pan Road was to observe whether throughput does actually increase or decrease as a result of very long greens and cycles. In the foregoing discussion, however, the existing data were used to suggest the throughput attainable using shorter cycles. This option was not available for field trial, but having a reasonable simulation made exploring the question possible.

Evaluating the improvement, however, requires a means of characterizing the operation of the intersection with respect to throughput. The simulation was repeated to compare the offered load (demand) to the served load (throughput) on the northbound approach for a range of input volumes. All simulation parameters were kept the same except input volume. Right turns were kept at the same percentage of the total approach traffic, which was 12.5%.

Offered load is the total input volume that VISSIM was asked to process that was routed onto the northbound through lanes (right turners are excluded in order to provide a comparison to field data, which measured only through cars on a cycle-by-cycle basis). Because the residual queue spilled out of the simulation network (the size of which was constrained by the limited VISSIM license), it was not possible to characterize demand upstream from the queue. Thus, it was necessary to determine whether VISSIM's stochastic traffic generator followed sufficiently close to the nominal traffic volumes to allow the use of the nominal values in the evaluation. A comparison was therefore made between nominal volume as programmed in the VISSIM input file and actual volume presented to the approach in the simulation to determine how subject the value was to stochastic processes within the simulation. The numbers were found to be within a few percentage points in all cases where the residual queue did not spill out of the network. Thus, the offered load was assumed to be the hourly demand that was nominally programmed into the simulation.

Table 3-1 shows the relationship between nominal programmed demand and volume measured at the upstream end of the entry links. For values up to the maximum throughput, the nominal demand and the actual demand were very close.

Nominal Programmed Demand (vph)	Actual Simulated Demand (vph)	Error (%)
1000	1016	1.6
1500	1473	1.8
2000	2024	1.2
2500	2468	1.3
3000	2959	1.4
3500	3471	0.8
4000	3979	0.5
5000	4008	19.8
6000	4019	33.0

Table 3-1. Nominal and Actual Simulated Volumes

If demand sufficiently exceeded the maximum throughput at the stop line, the approach would overflow, but VISSIM would still be attempting to feed vehicles into the network at the programmed rate. The served throughput was the total volume of traffic measured at data collection points in the through lanes at the stop line, as recorded from VISSIM.

Figure 3-10 shows offered versus served loads at the simulated intersection. For an intersection that is able to serve all the demand presented to it, the chart should show a one-to-one relationship between offered and served loads. When the intersection reaches the point where additional loads cannot be served, the curve relating the two flattens to a horizontal line that represents maximum throughput.

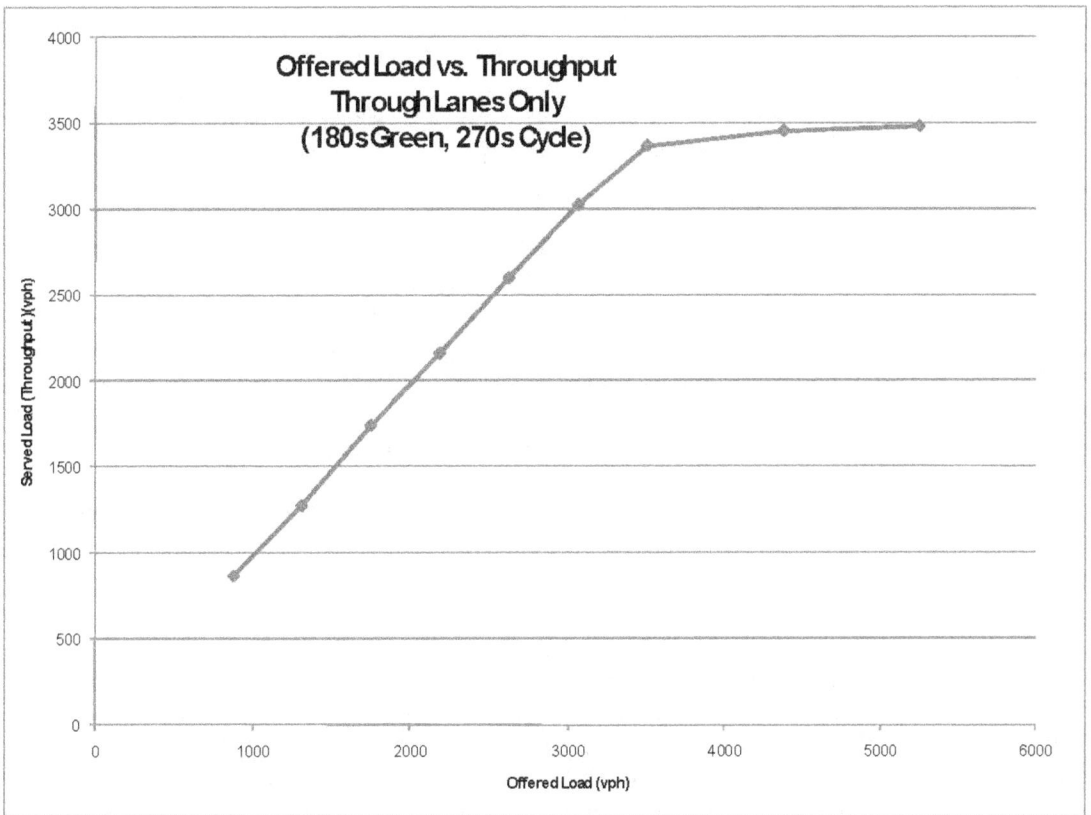

Figure 3-10. Offered Load vs. Throughput, from Simulation

The figure shows that throughput tracks demand in the absence of growing residual queues, as expected, until the maximum throughput is reached. At that point, additional demand cannot be accommodated and the served throughput reaches a ceiling. In the simulation, the maximum throughput was a little less than 3500 vehicles per hour, which corresponded reasonably well to the maximum observed throughput of about 3300 vehicles per hour reported in the field data. The difference can be explained by the slightly reduced headways as simulated versus observed.

This approach in identifying the maximum throughput by comparing offered and served load provides a means of evaluating alternative strategies for minimizing congestion on the basis of the throughput objective. For this site, the main alternative under consideration is related to the assumption that increased cycle lengths increase throughput, so the curve in Figure 3-10 will be reproduced at several cycle lengths to determine if cycles affect maximum throughput.

The simulations were repeated at two other sets of signal timing, to correspond to the thought experiment in the discussion of the field results. In addition to the field case of a 180-second green within a 270-second cycle, simulations were conducted with a 48-second green in a 72-second cycle, and a 118-second green in a 177-second cycle. Cycle-by-cycle summary data were collected for each case.

It should be noted again that the 72-second cycle is impractically short for an intersection such as this one. It is included because the green time matches the time required to clear the queue back to the upstream end of the right-turn bay. Even so, including it in this evaluation is *reductio ad absurdum*. At the 72-second cycle, the side-street and opposing left-turn movements would be ridiculously short (about 3 seconds of green). In the simulation, however, the 3-second greens were able to serve typically two cars in each lane on the minor movements per cycle. The short cycle comes more often, and, with 50 cycles per hour, the two-lane minor movements would be expected to serve around 200 vehicles per hour each. The demand exceed these values, but not overwhelmingly so. Even in this rather extreme case, the minor movements might leave 300 vehicles queued at the end of an hour, with approach volumes on the side street of about 300 and about 400 on the left turn opposing the movement under consideration. But northbound through traffic typically must wait through three cycles to clear the intersection, and this was experienced by the data collector on the morning the field study. With a 180-second green, three cycles of delay means over 800 vehicles in the queue. Even blocking the minor movements completely would create less congestion than that. Of course, the public would not accept such draconian measures, let alone be able to maintain safe operation. And public pressure would compel the practitioner to increase the green time on the minor movements, which would reduce the percentage of the cycle given to the congested through movement. That would substantially decrease throughput.

The 177-second cycle, on the other hand, provides green times for the minor movements of 21 seconds, which are long enough to provide adequate minimum greens even at an intersection with demand as lopsided in favor of one movement as this one. In the simulation of the 177-second cycle, there was no congestion on the other movements, and the regular queues were shorter than for the 270-second cycle, as would be expected. Minor-movement motorists would see a noticeable reduction in delay.

Figure 3-11 shows the maximum throughput curves for the three timing scenarios.

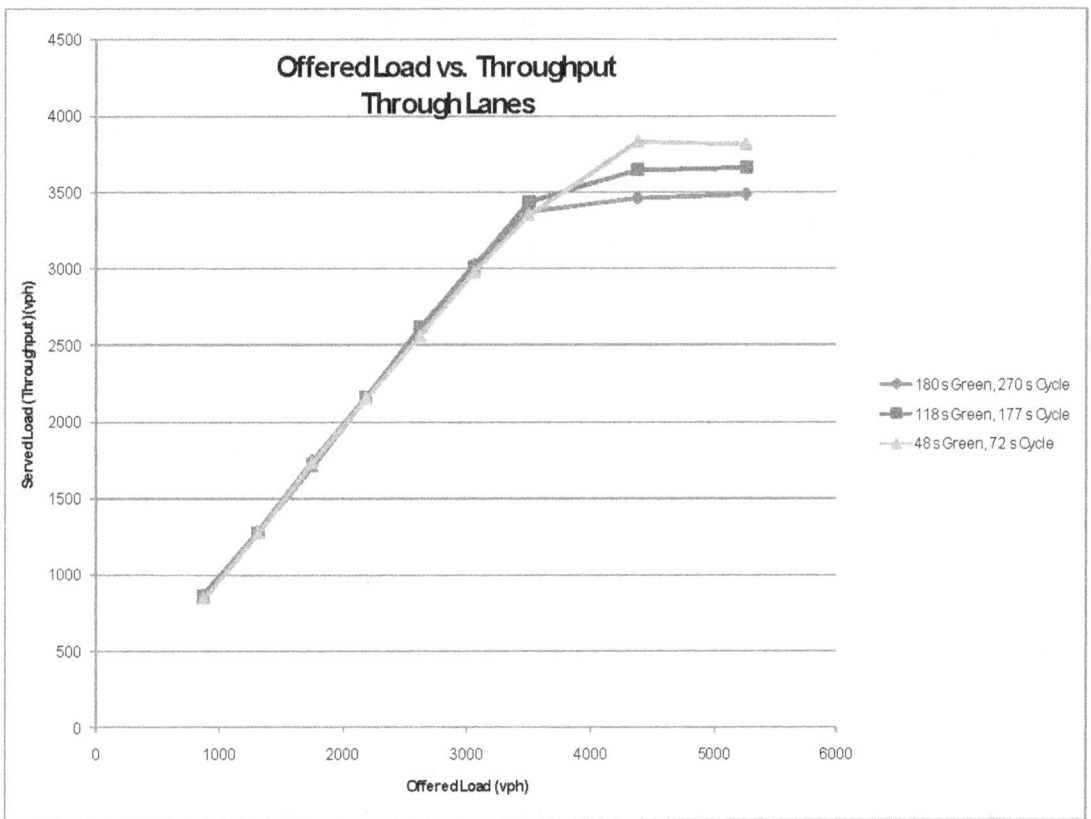

Figure 3-11. Simulated Maximum Throughput for Three Signal Timing Scenarios

Discussion and Conclusion

Two influences on the throughput compete against one another. The first is the effect of phase-change lost time. The clearance intervals were set to six seconds in all timing scenarios, and the three conflicting signal phases in the intersection each contributed six seconds of clearance intervals. Lost time is usually assumed to be about four seconds, but again it is constant. The clearance intervals are only 6.7% of the 270-second cycle, but are a full 25% of the 72-second cycle. Lost time is 4.4% of the longer cycle and 17% of the short cycle. As was mentioned at the outset, the consumption of the cycle by lost time is the usual motivation for using longer cycles in the presence of capacity problems and congestion, based on the assumption that the longer cycle is less consumed by lost time and therefore more efficient.

The competing influence is the effect of the right turn. As we saw in both the field data and the simulation, the headways and saturation flow of the right lane was similar to the middle and left lanes, and also fairly constant, until the queue emptied to a point adjacent to the right-turn bay entrance. In that section of roadway, all the vehicles in the right lane are through vehicles, because right turners have already moved into the right-turn lane. Upstream from the right-turn lane entrance, however, the right through lane contains all the right turners who are unable to reach the right turn lane because of the residual queue. Once the queue clears back to the point where right turners are present in the queue, their departure into the right-turn lane as the queue clears will leave gaps in the through traffic. Through cars attempt to fill the gap, partly from further back and partly from the middle lane, and a shock wave develops. Headways spike, recover, but then spike again as the shock wave crossing the stop line, as was seen in both the field data and the simulation.

At the field site, the effect of the right turn was sufficiently dominant to have a significant effect on the overall average headway on the approach.

Looking in the field data at the number of vehicles served, on average, across all three lanes at three sample green times of 48 seconds, 118 seconds, and 180 seconds, and then converted to hourly flows, showed that all three timing scenarios provided about the same throughput.

Given that the 48-second green time serves only cars queued downstream of the entrance to the right turn lane, we expect the headways to be uniformly short and unaffected by right turners. The simulation showed this by indicating that the 48-second green time provided the *highest* maximum throughput, assuming that the green percentage of the cycle was held constant.

The simulation also indicates that the larger the percentage of the green that can be consumed by flows unaffected by turning traffic, the higher the throughput. The medium cycle in the simulation provided higher throughput than the longest cycle, but not as high as the shortest cycle. This result is the opposite of the assumption made by most practitioners, and indicates that in this simulated case, at least, the effect of turning traffic on overall throughput overcomes the reduced cycle efficiency caused by lost time. The simulation was perhaps more optimistic than the field data, though the field data represented only the longest cycle, and was manipulated to suggest what might be the performance of shorter cycles. In any case, a safe conclusion is that at this site, the longer cycle is not increasing throughput.

In conclusion, the field data did not support the hypothesis that lost time builds back into the traffic stream as queues empty through very long greens. But turning traffic leaves gaps in the through lanes, and the data at this site show that those gaps *do* result in a significant decrease in throughput. Simulation of the intersection suggests that decrease is more pronounced the when the green is long enough such that a significant portion of it serves queue containing turning vehicles. The less green time is used to serve a queue containing turning vehicles, the greater the throughput in the through lanes.

The problem with short cycles is not that they are inefficient, but that they run afoul of external complaints and constraints that result in green time being distributed unfairly to minor movements. Thus, the study of this site confirms the recommendation that emerged in the state of the practice to find the *right* cycle, which in this case should be *just long enough to allow green times to be assigned equitably based on traffic demand.*

Cycle Length and Bus Capacity

Introduction

One of the strategies that emerged when researching the state of the practice was the relationship between bus flow and cycle length. The scenario described was in a tight grid network, such as a central business district, where a cycle that did not resonate with the normal stop time of buses causes those buses to build a queue, which in turn propagated into the general traffic stream. This case was observed some years ago in the San Antonio business district, particularly on Commerce Street, which is a one-way street in a tight grid network. A very popular bus stop on Commerce at Soledad served high bus volumes of about 85 per hour. With the 100-second cycles that were existing at the time, buses frequently queue through upstream intersections. A change in cycle length to 60 seconds (which was general resonant in most portions of that network) eliminated these bus queues. The resulting experience suggests that the capacity of bus traffic through an intersection is one bus per cycle, and with shorter cycles, more buses per hour can be served. The researchers investigated this suggestion using simulation.

The simulated scenario used the section of Commerce Street in downtown San Antonio where the original observations had been made.

In the aerial photo below, the intersection of Commerce and Soledad is in the upper left, and two upstream intersections are shown, including St. Marys and Navarro. Soledad is one-way northbound, St. Marys is one-way southbound, and Navarro is one-way northbound. Commerce extends to the right (east) from that upper-left intersection. The modeled network would include these three intersections. Commerce Street is one-way westbound. The location is not generally congested, but the bus flows approaching the stop on the Commerce approach to Soledad are quite high. In the simulation, the researchers used higher than realistic volumes to create congestion. (The photo shows a bus at this stop, and two more turning into Commerce from southbound St. Marys).

Figure 3-12. Commerce Street at Soledad, St. Marys, and Navarro, San Antonio

Modeled Operation

Signal Operation

The network was configured with three pre-time controllers. In all cases, the intersections, which are about 500 feet apart, were set up with offset of 12 seconds behind the preceding intersection. Thus, the green at St. Marys started 12 seconds following the start of green at Navarro, and the green at Soledad started 12 seconds later still. This provided perfect one-way progression along Commerce Street at a speed of a little under 30 mph.

To meter traffic into the test network, the green time (including clearance) along Commerce at Navarro was constrained to 50% of the cycle. Commerce was given 80% of the cycle at St. Marys and at Soledad. With the first intersection metering traffic into the network, any change in queuing upstream from that point would be caused by a significant effect at the two downstream intersections. In other words, the constrained green time at the first intersection reduced the sensitivity of the queue formation upstream from that point to changes in the throughput at the downstream locations. This had the effect of filtering the noise of insignificant effects.

Cycle lengths were varied over a wide range, from 25 seconds (which is acknowledged to be unfeasibly short because it violates pedestrian crossing requirements) to 100 seconds.

Demand

A base condition of 5200 vehicles per hour formed the simulated volume in the study. This volume caused significant congestion, the extent of which was the object of measurement The volumes are shown in the table below:

Entry Point	Volume
WB Commerce at Navarro	3000
NB Navarro	400
SB St. Marys	600
NB Soledad	1200
Total	5200

Table 3-2. Traffic Volume Scenario

Bus demand was modeled in a range of conditions, varying from the base bus demand of 50 buses/hour entering on St. Marys and 35 buses/hour entering on Commerce to 120% of those values. Bus stops were modeled with a 20-second stop period for passenger loading and unloading.

Simulation

In VISSIM, Commerce street was modeled with a very long external approach to provide a large storage space for measuring the extent of queue formation. All simulated scenarios were replicated for five simulation runs. Figure 3-13 shows the VISSIM network in the vicinity of the signalized intersections, and Figure 3-14 shows the overall VISSIM network with the long approach for storing the residual queue.

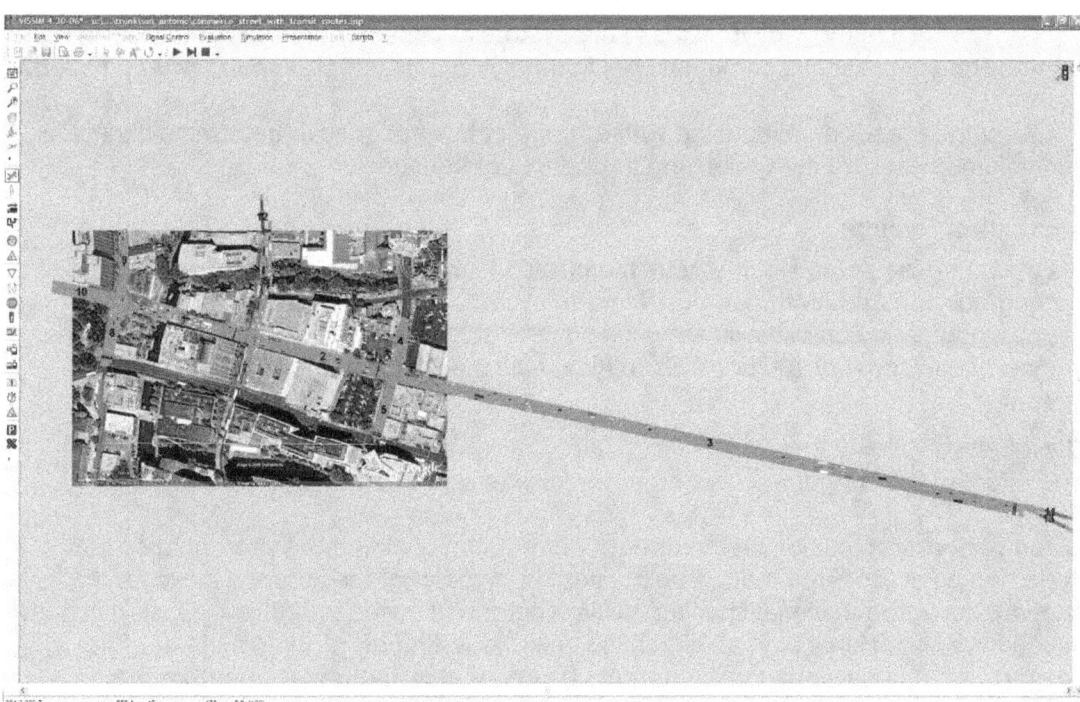

Figure 3-13. Entire VISSIM Network, Including Queue Storage

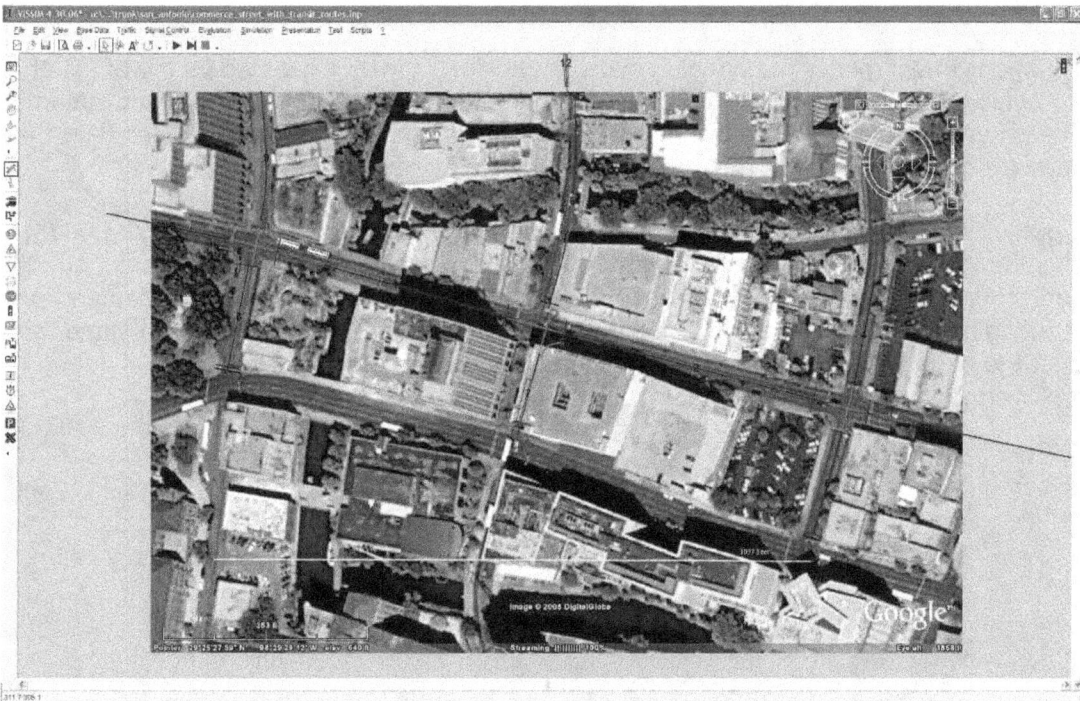

Figure 3-14 VISSIM Network Showing Signalized Intersections and Links

In the simulation, the queues built for the first hour of simulation, and the demand was dropped to a small fraction of the original demand to determine how quickly the network would be able to empty the queue that had been formed. Two results were therefore anticipated:

- The total number of vehicles in the system in one hour of traffic demand
- The effectiveness of the network in emptying traffic after demand was removed.

Additionally, a test was conducted with a range of bus volumes to determine if varying bus volumes affected the underlying throughput of the street.

Simulation Results

In comparing the served load versus the offered load no change was noted as a result of varying bus demand. With non-bus demand at 5200 total vehicles entering the network over an hour, and bus demand varying to 120% of the base condition. This was at the longest cycle length of 100 seconds to maximize the effect of buses on the traffic stream.

Cycle Length
The effect of cycle length, on the other hand, was significant.

Given perfect progression on the one-way street, the question of the resonance of the cycle has been rendered moot. The concept of cycle length resonance looks for the relationship between signal spacing, cycle length, and speed to find progression in both directions of an arterial or in all directions in a grid. With only a one-way street, the downstream intersections merely need to be delayed by the travel time from the upstream intersection to provide green time for every arriving car, assuming those cars are not impeded by buses. This is especially true given that traffic was metered into the network by constraining the green time at the first signal.

Thus, any effect related to changing cycle length can only be explained by the effect of the cycle length on the buses, and the resulting effect of the buses on the traffic stream. The research has already shown that variations in bus demand alone do not explain changes in the queue performance of the congested traffic stream.

Figure 3-x2 shows the accumulation of vehicles (i.e. the number of cars being served plus the residual queue) during the course of the simulated scenarios. Each point on each curve represents 60 seconds of simulated time. The point at which the demand is reduced is at 3600 seconds into the simulation. All scenarios recovered at the same rate after the demand was removed, though those that started with a larger residual queue took correspondingly longer to clear.

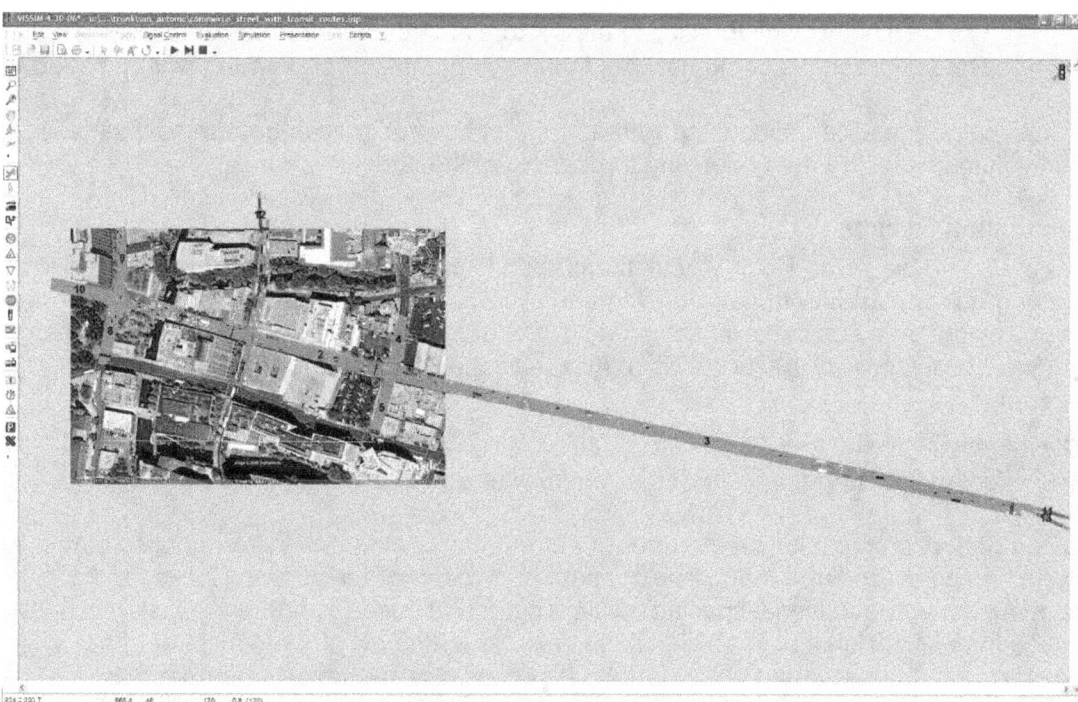

Figure 3-13. Entire VISSIM Network, Including Queue Storage

Figure 3-14 VISSIM Network Showing Signalized Intersections and Links

In the simulation, the queues built for the first hour of simulation, and the demand was dropped to a small fraction of the original demand to determine how quickly the network would be able to empty the queue that had been formed. Two results were therefore anticipated:

- The total number of vehicles in the system in one hour of traffic demand
- The effectiveness of the network in emptying traffic after demand was removed.

Additionally, a test was conducted with a range of bus volumes to determine if varying bus volumes affected the underlying throughput of the street.

Simulation Results

In comparing the served load versus the offered load no change was noted as a result of varying bus demand. With non-bus demand at 5200 total vehicles entering the network over an hour, and bus demand varying to 120% of the base condition. This was at the longest cycle length of 100 seconds to maximize the effect of buses on the traffic stream.

Cycle Length

The effect of cycle length, on the other hand, was significant.

Given perfect progression on the one-way street, the question of the resonance of the cycle has been rendered moot. The concept of cycle length resonance looks for the relationship between signal spacing, cycle length, and speed to find progression in both directions of an arterial or in all directions in a grid. With only a one-way street, the downstream intersections merely need to be delayed by the travel time from the upstream intersection to provide green time for every arriving car, assuming those cars are not impeded by buses. This is especially true given that traffic was metered into the network by constraining the green time at the first signal.

Thus, any effect related to changing cycle length can only be explained by the effect of the cycle length on the buses, and the resulting effect of the buses on the traffic stream. The research has already shown that variations in bus demand alone do not explain changes in the queue performance of the congested traffic stream.

Figure 3-x2 shows the accumulation of vehicles (i.e. the number of cars being served plus the residual queue) during the course of the simulated scenarios. Each point on each curve represents 60 seconds of simulated time. The point at which the demand is reduced is at 3600 seconds into the simulation. All scenarios recovered at the same rate after the demand was removed, though those that started with a larger residual queue took correspondingly longer to clear.

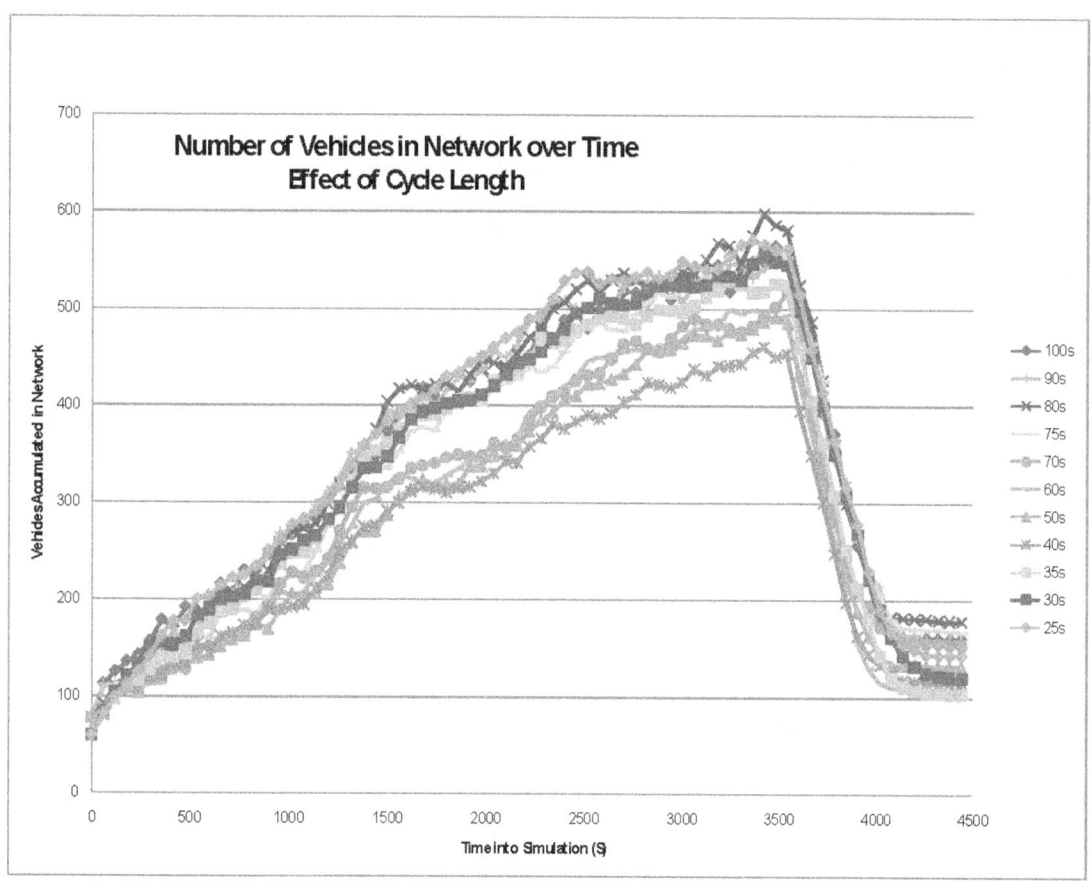

Figure 3-15. Effect of Cycle Length on Network Loading

The figure shows two important features. The first is that the cycles that stored the least number of vehicles were in the middle of the range, with the best scenario storing a maximum of about 450 cars at the end of the hour of loading, compared to about 600 vehicles for the worst cycle, for a one-third decrease in network loading. The cycle lengths ranging from 50 to 70 seconds was only a little worse than the best cycle, which was 40 seconds. That the 40-second cycle would perform best is plausible in addition to being observable: That cycle was the shortest (and therefore most frequent) that provided enough time to reliably load and unload one bus. Shorter cycles might require some buses to wait two cycles because their loading time extended beyond the available green, while much longer cycles might trap buses long enough such that another bus might have been able to process through the stop had the signal not been red so long.

The 80-second cycle showed the worst performance. For longer cycles, the green time could sometimes accommodate two buses, which to some extent offset the effect of fewer cycles per hour.

The other feature of the graph above is that the area under each curve represents the total delay in the network. Delay is measurable only because the queue is allowed to dissipate fully, and because the total time is the same for all scenarios. This would not be the case in most field situations without the ability to shut off demand completely to allow the queue to dissipate.

The area under each curve represents the total delay, and this can be converted to average delay by dividing this area by the total demand on the network during the simulation. Average delay is not meaningful in any specific case, because as the residual queue grows, delay grows with it. Average delay is therefore just a surrogate for the performance of that scenario rather than a model that can be used to predict what any one driver might see. Figure 3-X3 shows delay for each individual simulation run at each cycle length as a point, and the average of these replications as a curve.

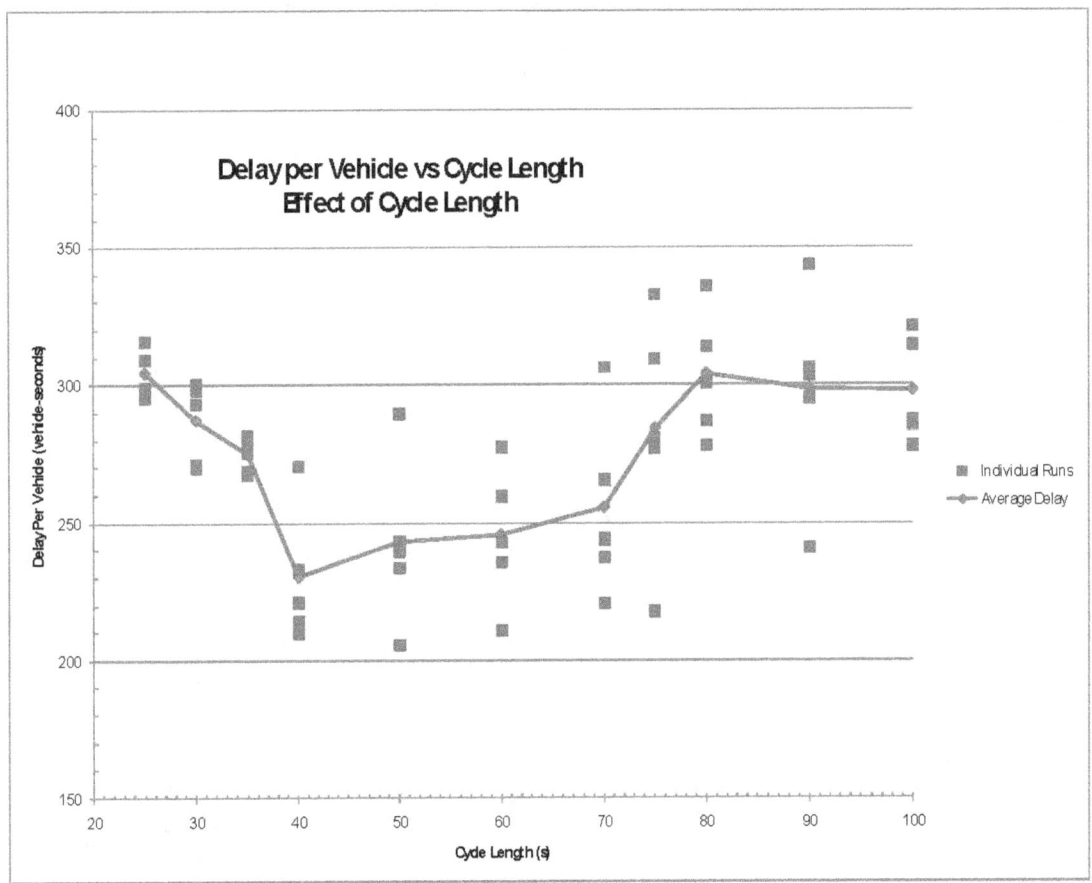

Figure 3-16. Average Delay Over a Range of Cycles

The figure reinforces the result that the 40 second cycle, which provides the most number of cycles per hour that can reliably serve at least one bus, performs the best. And the 80-second cycle, which is the longest cycle that can reliably serve *only* one bus, performs the worst.

Conclusion

In addition to providing a proper cycle length for progression resonance within a coordinated network, another resonance with bus loading times may also affect the cycle length decision. Cycles that are long enough to reliably serve one bus, but no longer, provide the best overall operation on congested routes that are affected by large bus volumes.

Bus volumes affect congested operation disproportionately to their numbers. A change in bus volumes has little effect on the congested network, but bus volumes that are only one or two percent of the total network flow can significantly affect the development of residual queues in a congested network.

www.ingramcontent.com/pod-product-compliance
Lightning Source LLC
Chambersburg PA
CBHW081843170526
45167CB00007B/2894